SPATIAL TECHNOLOGY AND ARCHAEOLOGY

The Archaeological Applications of GIS

SPATIAL TECHNOLOGY AND ARCHAEOLOGY

The Archaeological Applications of GIS

David Wheatley
University of Southampton

and

Mark Gillings
University of Leicester

Taylor & Francis
Taylor & Francis Group

Boca Raton London New York Singapore

A CRC title, part of the Taylor & Francis imprint, a member of the
Taylor & Francis Group, the academic division of T&F Informa plc.

First published 2002 by Taylor & Francis
11 New Fetter Lane, London EC4P 4EE

Simultaneously published in the USA and Canada
By Taylor & Francis Inc.
29 West 35th Street, New York, NY 10001

Taylor & Francis is an imprint of the Taylor & Francis Group

Every effort has been made to ensure that the advice and information in this
book is true and accurate at the time of going to press. However, neither the
publisher nor the authors can accept any legal responsibility or liability for
any errors or omissions that many be made. In the case of drug
administration, any medical procedure or the use of technical equipment
mentioned within this book, you are strongly advised to consult the
manufacturer's guidelines.

British Library Cataloguing in Publication Data
A catalogue record for this book is available from the British Library

Library of Congress Cataloging in Publication Data
A catalog record for this book has been requested

ISBN 0-415-24640-7 (pbk)

Rosina, Alice and Ellen (DW)

Cabes (MG)

Contents

List of figures

List of tables

Preface

This book is written for anybody who is involved with the interpretation or management of archaeological information, whether student, researcher or professional. It is, we hope, an accessible introduction to spatial technologies, the most ubiquitous of which are Geographic Information Systems. These have found widespread application to both archaeological research and management but, until now, archaeologists have had to rely either on introductions that are aimed at other disciplines (such as geography, ecology or soil science) or on the dispersed and often very specialised literature about the applications of GIS. That this literature is now so large speaks volumes about the interest in GIS within archaeology, but this should not really surprise us: archaeological information is so inherently *spatial*.

The book has its origins in the GIS courses that each author has taught at, respectively, Southampton and Leicester Universities and we draw considerably on our experience of how archaeologists most effectively learn about computer technologies in deciding how to present the material. Our approach has been to summarise the main methods and applications that most professionals and researchers will want or need to know about, while at the same time providing guidance to more detailed or advanced reading which can be selectively pursued. This is not a computer manual, so we do not guide the reader through which buttons to press on specific software systems. Instead, we try and explain what those buttons do, and how to combine software functions to do useful and interesting things. In this way, we intend that the reader should gain knowledge and skills that outlive the increasingly short life of individual software systems. Where appropriate, we give examples of how the methods we discussed are implemented in particular software products.

The areas we have chosen to include are a short account of how archaeology has thought about space, followed by an introduction to spatial databases and the basic ways that GIS manipulate spatial data. Because of their ubiquity and usefulness, we dedicate a whole chapter to the construction and use of elevation models. We then describe some of the major kinds of spatial analysis and cover the use of spatial technology for the analysis of sites, territories and distance; location and predictive modelling; trend surface and interpolation; visibility analysis and cultural resource management. Finally, we have indulged ourselves a little and allowed ourselves to imagine what directions the archaeological applications of spatial technologies will take in the future.

Inevitably, we have not been able to exhaustively cover all the areas that we would have wished. We would have liked to include far more about the advanced areas of GIS that we discuss in Chapter 12, for example, and to provide more detailed examples of many of the subjects covered. To do so, however, would have required at least another book of the same length as this one. The result, we hope, is a book that will provide 'one stop shopping' for the majority of researchers and professionals but without the pretence of 'completeness'. We hope that you find this book useful.

David Wheatley and Mark Gillings, June 2001

Acknowledgements

DW: John G Evans convinced me to do more than turn up to lectures and sparked my first interest in computing; Stephen Shennan encouraged my initial interest in GIS and Peter Ucko was badly enough advised to offer me a job. Moreover, this book owes much to the many MSc students I have taught at Southampton and also to Graeme Earl, David Ebert, Sennan Hennesy, Sarah Poppy, Tim Sly and James Steele who kindly commented on various sections. Many other friends and colleagues contributed to this book and particular thanks to Simon Keay, John Robb, Leonardo Garcia-Sanjuan, Doortje Van Hove, Quentin Mackie, Clive Gamble, Herb Maschner and Josh Pollard. Most of all, thanks to Rosina for her tireless reading, checking, bibliography entry and so much more.

MG: To Arnold Aspinall—a brilliant mentor who saw some worth and nurtured it and Steve Dockrill, who taught me everything useful I know. Finally, to Josh Pollard for first lending me a copy of *The Appropriation of Nature*.

We would both like to record our gratitude to the many organisations and individuals who have assisted us with the reproduction of copyrighted material or other information, including the GeoData Unit of the University of Southampton, Forest Enterprise (Wales), Hampshire County Council Environment Group, Space Imaging, IOS press, Alasdair Whittle, Nick Ryan and André Tschan. Particular thanks are also due to Martin Fowler for invaluable assistance with satellite images.

CHAPTER ONE
Archaeology, space and GIS

"One of the striking things about geographical information systems is the difficulty one has when one attempts to talk about them, or at least to talk about them in ways that are not strictly technical." (Curry 1998: xi)

This volume comprises a detailed examination of the relationship between archaeology and a particular group of computer-based applications. We use the term 'spatial technologies' to mean any technology concerned with the acquisition, storage or manipulation of spatial information, but by far the most widespread of these are Geographic, or Geographical, Information Systems, almost exclusively referred to by the acronym GIS. In less than ten years, GIS have progressed from exotic experimental tools, requiring costly workstations and only available to a few specialist researchers, to widely available technological platforms for the routine analysis of spatial information. Archaeology, in common with all disciplines concerned with the interpretation of geographically located material, has witnessed an unprecedented transformation of the methodological tools it uses for spatial records and analysis. Whereas a decade ago few grant-giving bodies or cultural resource managers would have recognised the term GIS, most would now be surprised to see a regional archaeological project that did not claim to utilise it.

Unfortunately, the growth in availability of GIS software has not always been accompanied by a corresponding increase in the knowledge and technical capabilities of archaeologists. All too often, archaeological research or cultural resource management projects begin with the vague idea of 'GIS-ing' the data and undertaking some 'GIS-ish' analysis with it later. When later arrives, it becomes clear that poor decisions taken in ignorance about recording have been compounded by a lack of understanding of the capacities and limitations of the technology. The inevitable result is, at best, analysis that fails to live up to expectations and, at worst, flawed data-sets, poorly documented resources and misleading conclusions.

That is the reason for this book. In writing it, we hope to provide archaeologists with a one-stop resource from which they can gain a far clearer picture of how GIS really can facilitate better archaeological research. We will also touch upon a number of broader issues, such as the wider role of spatial information in archaeological analysis and interpretation, and we will look at a number of related technologies concerned with the collection, management, analysis and presentation of such data. These include Computer-aided Mapping, Remote Sensing, Global Positioning Systems (GPS) and Terrain Modelling. Despite this breadth, it is important to stress that this is a book about GIS and archaeology, and as such aims to provide archaeologists with a detailed understanding of what GIS is, how it works, and how it can be critically and sensibly applied in everyday practice. To this end, care has been taken to ensure that the focus of the book is upon the relevance of GIS to archaeological research and management.

Where appropriate, individual software systems are mentioned, but the book is not intended as a tutorial for any particular suite of programs. In examining the relationship between archaeology and GIS our aim has been to identify generic problems and extract the salient features which are common to the majority of systems: opening up the 'black box' and carefully explaining what is going on inside. It is our firm belief that this is the only way to ensure that the knowledge gained will remain relevant and useful regardless of the specific software system with which the reader is currently engaged.

Until now, archaeologists who wish to understand GIS have been faced with a number of options, none of them particularly satisfactory or convenient. They could read introductory texts derived from other fields or disciplines, such as soil science or economic geography. Alternatively, the curious could be directed towards the proprietary manual accompanying a given piece of software and then decide how best to translate this information into a specific archaeological context. Although some manuals are excellent, this leaves the archaeologist to fashion an application of GIS that is sensitive to the unique requirements of the discipline, in the process adding an extra burden to the already non-trivial task of getting to grips with a complex technology.

Another possibility has been to seek out a number of specialised publications, often in rather obscure journals or proceedings and invariably concerned with only a small aspect of the subject matter. In recent years the appearance of a number of excellent edited volumes has served to mitigate this situation to some extent—examples include Allen, Green, and Zubrow (1990), Lock and Stancic (1995), Aldenderfer and Maschner (1996), Maschner (1996a), Johnson and North (1997), Lock (2000) and Westcott and Brandon (2000). However, these volumes are primarily aimed at researchers who already understand the technology, and so they rarely provide any obvious entry routes for the uninitiated. A good example is the growing literature on visibility analysis in archaeology. Almost without exception, the published papers on this subject treat the actual calculation of the viewshed as a given, leaving the archaeologist who comes to this literature with no easy point of access. It is inevitable that these kinds of publications will assume a level of technical literacy above and beyond the basics because the alternative is to flood the scientific literature with repeated and (to the expert) unnecessary detail. In the case of the example above, there is no simple answer as to how an archaeologist can gain knowledge of collecting topographical data, integrating it into the GIS, constructing a digital representation of the topography, and simulating what could be seen by a human observer standing at a fixed point upon it.

Those accounts that do explicitly consider the introduction of GIS into archaeology tend to be historical rather than foundational, charting the pattern of disciplinary adoption and firmly embedded within discussions of developmental trajectories. It is therefore clear that a significant gap exists and this book seeks to fill it. It assumes no previous familiarity with GIS, or similar systems, and is dedicated to illustrating the ways in which GIS can be applied to particular situations, highlighting those aspects of GIS that have proven most interesting or useful in this respect.

1.1 SPATIAL INFORMATION AND ARCHAEOLOGY

Archaeologists have long been aware of the importance of the spatial component of the archaeological record. Highly precise maps and ground-plans date back to the 18th century, and some of the earliest formal excavations are notable for the meticulous way in which the locations of finds were recorded. Take for example the work of the pioneering British archaeologist General Pitt-Rivers at the late Roman earthwork of Bokerly dyke on the border of Dorset and Wiltshire in the Southeast of England (Pitt-Rivers 1892). Here a series of scaled plans and sections were used to display clearly the three-dimensional locations of all artefacts and features.

Much, if not all, of the data that archaeologists recover is spatial in nature, or has an important spatial component. As a discipline archaeology routinely deals with enormous amounts of spatial data, varying in scale from the relative locations of archaeological sites upon a continental landmass, down to the positions of individual artefacts within excavated contexts. Artefacts, features, structures and sites, whether monument complexes, chance finds of individual objects, scatters of ploughsoil material or rigorously excavated structural and artefactual remains, are all found *somewhere*. As well as the position of the feature or artefact itself there may also be a series of *relationships* between the locations of features and artefacts, revealed by significant patterns and arrangements relative to other features and things.

These 'significant others' can be features of the environment (rivers or particular combinations of resources, for example); other archaeological features (such as hearths, storage pits, market-towns or ceremonial centres); or cosmological phenomena (symbolically significant rock outcrops, named places or the positions of the heavenly bodies). It does not take much thought to come up with a host of examples of this perceived patterning of artefact locations in space and the importance of knowing the spatial relationships between artefacts. A non-exhaustive list is presented in Table 1.1.

Some archaeologists such as David Clarke (1977a) even went as far as to claim that the retrieval of information from these spatial relationships should comprise the central aspect of the discipline. Archaeologists have long had an intuitive recognition of the importance hidden within spatial configurations and have expended considerable amounts of energy developing ever more efficient ways of recording it.

Over the last thirty years, the quality and volume of the spatial data we collect have increased dramatically as new surveying techniques and equipment have become routinely available to archaeologists. In the time it would have taken to measure and log two hundred locations using traditional tools such as a plane-table, a total station can log two thousand locations to sub-millimetre precision and recent developments in low-cost, high-precision Global Positioning Systems (GPS) are only serving to extend this trend. We will look at some of these techniques in more detail in Chapter 3.

Table 1.1 Some suggested spatial patterns and explanatory phenomena.

Spatial phenomenon	Explanatory factors
Seasonal Hunter-Gatherer camps within a landscape	The availability of resources
A hierarchy of settlements within a region	Territorial spacing and proximity to market
The distribution of watchtowers along a defensive structure (e.g. Hadrian's wall)	Carefully structured and formalised intention linked to a specific functional requirement
The location of coarse potsherds in the ploughsoil	The result of a number of quasi-random post-depositional forces
The position of a zone of decoration on the body of a ceramic vessel	The physical manifestation of a symbolic-ideological design
The arrangement of earthworks and stones at a Neolithic henge site	The unplanned, accumulated product of everyday ritual practice
The suggested zodiacal arrangements of monuments around the site of Glastonbury in Southeast Britain	The tendency of the human eye to see structures in apparently random patterns of dots
Discard of bone debris around a hearth	The functional requirements of bone-marrow extraction undertaken whilst sitting around a smoky fire
The Nazca lines and geo-glyphs	Processional routes and pathways

1.2 THINKING ABOUT SPACE

As archaeologists constructed new interpretative frameworks for the study of the past, the uses to which the carefully recorded spatial component of the archaeological dataset have been put have also changed. For much of the early 20th century, archaeology was dominated by what historians of the discipline have referred to variously as the Classificatory-Historical or Culture-History phase (Trigger 1989). Central to this was the idea that the remains recovered by archaeologists corresponded to the material manifestations of specific peoples, *i.e.* distinct ethnic groups. During this period, processes of change and the writing of histories were tied up with the notion of cultural diffusion, with historical change seen as the result of the spread of materials and ideas from a core zone out to new areas.

The spatial distribution of cultural traits was seen as critical to such an explanatory framework and in seeking to explore such spatial phenomena, archaeologists and anthropologists on both sides of the Atlantic found an important source of inspiration in the work of geographers. In particular the Austro-German school of anthropo-geographers, typified by the work of Ratzel and later Boas (Trigger 1989:151-152). One of the aims of their work was to show how processes of diffusion created discrete 'culture zones' which corresponded to culturally homogeneous complexes which were clearly and unambiguously bounded in space (Clarke 1977a:2, Gosden 1999:70). These culture-areas were identified through the establishment of cross-classified trait lists based upon aspects of material culture, and then visualised through the careful plotting of the distribution of these traits upon maps. Once plotted, the spatial extent of culture-area and relationships between them could be visually assessed.

To give an example, artefacts (usually deriving from burial deposits) were classified into complex typological sequences and the occurrences of unique types and combinations of types were equated directly with the presence of distinctive groups of peoples—variously termed cultures, foci or traditions. Once defined on typological grounds, the spatial occurrences of the diagnostic types could be plotted and a line drawn around them to indicate the spatial extent of the culture group. By placing the artefacts in a typological sequence the culture could be bounded in time as well, and changing configurations and relationships between these culture zones examined (Aldenderfer 1996). These were then used to study and characterise the cultural patterns of a given region through time, reconstructing its cultural history in terms of diffusion, direct contact and acculturation.

As well as a concern with the characterisation and spatial delineation of 'cultures', there was, in both Europe and the US, a concomitant concern with the mapping of prehistoric settlements against environmental variables. This activity was encouraged by the work of the anthropo-geographers. In England such developments served to reinforce an existing emphasis upon the role of landscape and geography as conditioning factors in the form of historic settlement patterns, leading to the work in the 1910s and 1920s of researchers such as Crawford, and later Fox. They used a series of archaeological and environmental distribution maps to communicate and explore change in a region through time (Trigger 1989:249).

In the 1940s interest in spatiality declined within British Archaeology in tandem with a reorientation away from diffusionist modes of explanation towards more economic and evolutionary schemes with direct emphasis upon spatial patterning. In the US the study of patterns and environment continued to develop throughout the 20th century, finding its clearest expression in the 'ecological' approach advocated by Steward in the 1920s and 1930s, and later elaborated by Willey (Clarke 1977b:3, Trigger 1989:279-285). Here the mapping of archaeological sites was undertaken on a regional scale, with the express purpose of studying the adaptation of social and settlement patterns within an environmental context *i.e.* to find causal linkages.

In such studies the main techniques used to explore and interpret the spatial component of the archaeological record were intuitive—usually the simple visual examination of distribution maps. In the distribution map, the relative spatial positions of things (in this case the objects we are interested in) are represented as a scatter of symbols spread out across a flat two-dimensional plan. This form of representation developed in England in the mid 19th century as a means of displaying and exploring the occurrence of disease outbreaks, principally cholera (Thrower 1972:94). The use of distribution maps in archaeology was popularised by individuals such as Cyril Fox and through the 'Field Archaeology' practised by the Ordnance Survey of Great Britain's first archaeological officer, OGS Crawford. Both recognised that the past could not be fully understood simply from the examination and characterisation of de-contextualised artefacts, or from the study of the sequences derived from individual sites in isolation. They were both explicitly concerned with recording and analysing the spatial dimension of archaeological material. The importance of such maps to the discipline was emphasised by Daniel who described them as one of the main instruments of archaeological research and exposition (Daniel 1962 quoted in Hodder and Orton 1976:1).

1.3 NEUTRAL SPACE AND QUANTIFICATION

It is fair to say that for the majority of the 20th century spatial archaeological data has been tabulated and plotted by hand on simple, flat maps. The analysis and synthesis of the carefully recorded spatial information was restricted to the visual appraisal of these static distribution maps, looking for similarities, trends and differences.

The early 1960s, however, witnessed a dramatic change in both the status afforded to spatial information and the techniques used to identify and explore spatial patterns and relationships. This was embedded in a much broader change in interpretative emphasis, which moved away from regarding material culture as a direct and non-problematic reflection of a distinct people: the unambiguous spatial footprint of some past culture. Instead, material culture was seen as resulting from a series of past processes and spatial relationships were seen as the spatial impact of these behavioural activities and processes. Within this new conceptual framework, earlier approaches to the study of spatial phenomenon were criticised for being limited in their aims and based upon methods that were often highly uncritical. The casual visual examination of distribution maps was seen as highly subjective and, as a result, potentially dangerous (Hodder and Orton 1976:4, Clarke 1977a:5). The more information contained in a given map, the more ambiguity and uncertainty there was likely to be in its interpretation.

With the quantitative revolution heralded by the New Archaeology of the 1960s, came the first real attempts to move beyond visual appraisal and explore in more detail the shape, form and nature of the spatial patterns visible in the archaeological record. If such patterns were to be studied it had to be in a more objective and verifiable manner. From this point on it was no longer sufficient to say that the locations of a number of features, for example burial mounds in a funerary landscape, *appeared* to be grouped together or that some *looked* to be aligned upon a particular landscape feature. Was there really a grouping or preferred alignment, and if so why? Rather than mere description what was needed was *explanation*.

Space acted as a canvas upon which cultural activity left traces. For example, rather than being regarded as distinctive 'peoples' or 'ethnic groupings', cultures were portrayed as systems, composed of distinctive sub-systems. In such systems stability was the norm, with change resisted by feedback between the various component units. Culture change was always the result of external factors (usually precipitated by changes in the ecological setting of any given system). As a result, variability in the spatial distribution of artefacts and structures was not explained exclusively in terms of different ethnic groupings, but instead as resulting from specific functional roles linked to the maintenance of stability in the face of external change. Environment was the prime external factor that was assumed to have affected the distribution of people in an otherwise homogeneous space, influencing, guiding or dictating behaviour and thus leading to the observable patterns recorded by the archaeologist. In adapting to changes, and limiting factors in the environment, human groups behaved according to utilitarian and functional aspects of living: minimising energy costs and optimising efficiency by maximising proximity to good soils, a market, or adopting the best defensive position. The behaviour of past peoples left clear patterns inscribed upon space, which archaeologists could subsequently identify and measure, thus providing them with a clear link back to the original causal factors.

The key point to emphasise is that external factors influenced behaviour, and this behaviour left patterns in space that could be objectively measured and quantified. Once measured, this information could be rigorously analysed and the effects of factors such as distance and the varying potentialities of site locations could be objectively specified (Tilley 1994:9). In this way it would ultimately be possible to identify the generative process. This new approach to the analysis and study of spatial data came largely through the rigorous application of a wide range of spatial analytic techniques and methods. The distribution map did not disappear, but it could be argued that its role was slightly altered. Rather than acting as the sole data layer upon which interpretations were based, it became instead a foundational data summary or stepping-stone for further detailed analyses.

To return to our earlier example, to determine whether a group of burial mounds were in fact grouped or clustered, as a first stage they could be surveyed and their locations plotted on a distribution map. Traditionally the archaeologist would simply look for likely clusters of points. In the new approach instead of simply saying that the burial mounds *appeared* to be grouped, you would begin with the hypothesis that they were not (termed a 'null hypothesis') and would test the statistical significance of that interpretation against your group of artefacts using, perhaps, a quadrat or nearest neighbour test (we will discuss such approaches in more detail in Chapter 6). This scientific approach of 'hypothetico-deductive reasoning' is characteristic of the set of approaches labelled New (or more recently 'processual') Archaeology. In a nutshell, emphasis and value was placed upon formulating hypotheses and then testing them against the data. If the hypothesis was rejected, it was modified accordingly and tested again. This can be contrasted to the inductive approach that preceded it whereby the data—most frequently in the form of a distribution map—was examined for patterns, on the basis of which hypotheses and interpretations could then be constructed. A number of statistical techniques were appropriated and developed to define and explore the significance of artefact distributions. These borrowings came largely from the fields of Geography, Economics and Plant Ecology.

As well as concerns with objectivity and rigour, there was also an important element of contingency behind this move towards quantification. As intimated in the introduction, the volume of spatial data collected by archaeologists was growing dramatically. As a result, new methods for the management and exploration of this swelling database were seen as not merely desirable, but essential.

1.4 MEANINGFUL SPACES

The spatial-scientific approaches characteristic of the New Archaeology, and the more impressionistic visual techniques that preceded them, shared one important factor in common. They were based upon a very fixed understanding of what *space* itself actually was. Space was viewed as a neutral, abstract dimension in which human action took place. It was a universal, clearly measurable, and fundamentally external backdrop to cultural activity. Space was a taken-for-granted category that, in itself, was non-problematic. The space of the Neolithic period was therefore identical to the space of the Medieval period, which was in turn identical to the space of the modern archaeologists.

Since the mid 1980s this notion of space as a non-problematic, abstract backdrop has begun to be increasingly challenged and questioned. This has come in the context of yet another change in the interpretative frameworks employed by archaeologists in the study of the past. These can be characterised by their criticism of approaches that sought to deny past peoples their essential humanity, for example viewing social groupings as 'systems' responding solely to external environmental stimuli. In such studies, broadly termed post-processual, approaches to spatiality have been informed by disciplines such as Ethnography, Cultural and Social Anthropology, Philosophy and Sociology. In such formulations, emphasis has been given over to notions of space that are intimately tied to cultural activity. Far from being seen as a uniform, abstract backdrop to social actions, space is seen as embedded and implicated in those actions. In the words of the prehistorian Chris Tilley "...space cannot exist apart from the events and activities within which it is implicated" (Tilley 1994:10). Rather than 'always being there', and always *having* been there, space is constructed and shaped by social actions which it in turn serves to construct and shape. The space of the Neolithic, the Roman period and the Medieval are as distinct and different from each other as they are from the space of the modern archaeologist.

Many archaeologists and anthropologists have begun to argue that space should not be seen as a neutral container for human action, but instead as a meaningful medium for human action. This has resulted in a questioning of the approaches to spatial patterning pioneered by the New Archaeology and the assumption that archaeologists could simply read patterns from space and reconstruct past activities. This has led to the development of much more embedded notions of dwelling and being-in-the-world, and the idea of spaces socially rendered through routine, everyday practice into arrangements of locales and places linked by pathways. Factors such as perception and movement have been brought to the fore as archaeologists have begun to explore and conceptualise spatial configurations in new and provocative ways. To quote the social anthropologist Tim Ingold "To proceed we need a set of terms to describe the geometry not of abstract, isotropic space, but of the substantial environment in which humans and other animals move, perceive and behave" (Ingold 1986:147).

1.5 WHAT IS A GIS?

Geographic Information Systems are a relatively new technology but despite their recent appearance they are now widely used in a huge variety of disciplinary contexts. They are also to be found in a wide range of practical applications, from forestry and sewerage management through to flood control. Unfortunately, the label 'GIS' is rather slippery and difficult to define in a precise or meaningful way. Generic definitions, as we will see shortly, are relatively straightforward to construct but are often so generic as to be almost meaningless. The reason for this elusiveness is that as individual disciplines have adopted the technology they have tended to tailor it to their specific requirements. In a sense the precise definition afforded to GIS is as much dictated by the end-use as any intrinsic characteristics of the systems themselves. Planners, geographers, economists and utilities companies—not to mention archaeologists—can offer markedly different definitions, which are equally valid yet differ in the specific stresses and emphases they place upon the functionality of the software.

There is also very little consensus on what actually defines a GIS. Some have even argued that the term is not at all meaningful or useful, but this is an academic point rather than a practical one. For our purposes, GIS software exists and performs tasks which many researchers, including archaeologists, have found to be extremely useful in managing and analysing their data. Such confusion has not, of course, prevented a great many people from trying to provide a definition of GIS. Without getting too introspective, it is perhaps worth citing the definitions offered by two widely read textbooks:

"... a powerful set of tools for collecting, storing, retrieving at will, transforming, and displaying spatial data from the real world for a particular set of purposes." (Burrough 1986)

"An information system that is designed to work with data referenced by spatial or geographic co-ordinates. In other words a GIS is both a database system with specific capabilities for spatially-referenced data as well as a set of operations for working (analysis) with the data." (Star and Estes 1990)

GIS, then, are computer systems whose main purpose is to store, manipulate, analyse and present information about geographic space. Of course, this is a definition which could apply to many technologies, for example computerised databases or even paper maps. This is because the individual components which make up GIS are not very new. In fact, they have been available for some considerable time as, for example, computer-aided cartography, computer-aided design (CAD), image processing and database management systems (DBMS).

It is important to acknowledge that a GIS is not a single, monolithic computer program, and a useful conception of GIS is as a 'spatial toolbox'. It comprises a combination of several different software technologies. Although conceptually neat, it should be noted that the tool analogy has provoked much heated debate amongst practitioners, some of whom see GIS more as a science than the mere application of a tool and others who question the theoretical neutrality such a term implies.

In practice, this also means that there are a wide variety of GIS. Some were written from first principles, specifically as GIS, while others are implemented as extensions to existing software. For example, adding a map-based graphical interface and some analysis routines to a relational database system can result in a

Figure 1.1 Contrary to popular mythology, contemporary professional archaeologists may spend more time using a GIS than a trowel. Ian Wykes (Senior Archaeologist, Hampshire County Council) reviews records of crop-mark sites.

minimal GIS. Database extensions to computer-aided design or cartography packages can provide very similar functionality. As a result of this, software systems that are referred to as GIS are extremely varied, and the boundaries between GIS, remote sensing, computer-aided mapping and database management are becoming increasingly blurred.

Given the emphasis afforded to spatial information by the GIS, it is crucially important to realise that there is more than one way of thinking about space. The concept of 'space' itself is far from unproblematic. To some space is a neutral backdrop upon which actions leave traces, always present and always the same. To other archaeologists space is a meaningful medium which is socially constructed, and within which actions are embedded, constituted in different ways and is always changing and it was into this complex existing tradition of interest in the spatial structure of cultural remains that the technology of GIS was imported in the mid 1980s.

1.6 AN ANATOMY LESSON

The various constituent programmes contained within the GIS can be grouped into clear application areas. Marble (1990) considers that GIS comprise four major subsystems:

1. The *Data Entry* subsystem handles all of the tasks involved in the translation of raw or partially processed spatial data into an input stream of known and carefully controlled characteristics.
2. The *Spatial Database* corresponds to Marble's Data Storage and Retrieval subsystems. It is responsible for storing spatial, topological and attribute information, and maintaining links with external database systems.
3. The *Manipulation and Analysis* subsystem takes care of all data transformations, and carries out spatial analysis and modelling functions.
4. The *Visualisation and Reporting* subsystem returns the results of queries and analyses to the user in the form of maps and other graphics as well as text.

To these four main subsystems, we might add the *User Interface*. Although not itself part of the flow of data, this is an essential component of a modern GIS because it is through this that users submit instructions to the other subsystems and obtain feedback on the progress of commands. The interface may be the familiar combination of screen and keyboard; a digitiser and puck; or even a simple web-browser and the quality of the interface to a great extent determines the 'learning curve' of the system. A good interface minimises the time needed for new users to become familiar with the basics, and at the same time ensures that experienced users can perform advanced functions rapidly (something we return to in Chapter 11).

Figure 1.2 shows the logical organisation of these components, together with the flow of data and the flow of control throughout the system. Input devices will usually include physical devices, but also programs that convert data from an

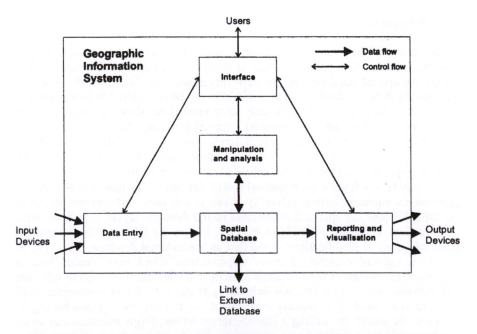

Figure 1.2 The major subsystems of a GIS as identified by Marble (1990) but adding the interface, inputs and outputs.

external to an internal data format—usually called import functions or import filters. Input devices can be broadly split into *control* devices, such as the keyboard and mouse, whose primary function is to send instructions to the system and *data* devices, such as digitising tablets and scanners, whose function is to obtain raw data. Output devices will either be transient, such as a video display screen, or permanent output devices such as a printer or pen plotter. A similar distinction between control and data devices might be made for output devices, although the most common form of output (the screen or 'video display') usually performs both control and display functions.

Data entry

Data entry tools enable us to get spatial information into the GIS. The most widely recognised element of the data entry system is the digitising program. This allows paper records to be correctly registered and transferred to a digital format (shown in Figure 1.2). We will look in detail at the process of digitising in Chapter 3 but digitising represents only one means of getting information into the GIS.

Other methods of data entry are also possible: maps can be scanned with an image scanner and then imported in raster format; lists of co-ordinates can be read from national and regional monuments registers; the data generated by survey instruments such as Total stations and Global Positioning Systems (GPS) can be read by the GIS; and pre-existing digital data (*e.g.* excavation plans stored in CAD) can be directly imported.

Spatial database

As this is one of the key areas that most distinguishes GIS from other mapping or drawing systems, the spatial database is discussed in detail in Chapter 2. Data within the spatial database are generally organised into *layers,* each of which represents a different theme. A spatial database can be considered to be similar to a collection of thematic maps, with each component sheet showing some different characteristic of the study area. Whereas a typical map may show roads, rivers and contours all on the same sheet, this is rarely the case in a GIS. In the spatial database such a map would be split into separate layers with one layer for the contour map, one layer for the river network and one for the roads and so on.

Many GIS will allow their internal spatial database to be linked to an external database management system (DBMS). When this is done, information stored in the DBMS *about* the spatial features held in the thematic layers, which we term attribute data, are made available to the GIS. In turn the GIS can also be used to make additions or modifications to the data in the external database. This type of facility is especially useful in larger organisations which may already have a substantial investment in their database system, and who therefore wish to integrate GIS into their existing information technology strategy. So far as manufacturers of GIS are concerned, it is equally convenient, as it frees them from having to 'reinvent the wheel' by writing a DBMS, report writer, forms interface and so on for their systems.

The visualisation and reporting system is frequently bound up closely with the interface to the GIS. This is because the most intuitive method of interacting with spatial data, and constructing queries, is often to simply 'point' at a visual representation of a map. GIS all provide functions for the creation of maps on screen, allowing the user to specify which data layers are displayed and how they are to be represented. In addition, many systems provide functions for producing printed maps, Internet-ready graphics files and alternative methods of representing the data on screen or paper. Three-dimensional net or block diagrams are particularly popular alternatives and are discussed in more detail in Chapter 5. Thanks to the reporting and visualisation system it is also a straightforward task to extract textual reports, summary statistics and numeric listings from the systems.

Manipulation and analysis

The manipulation and analysis subsystem generally comprises many different functions for performing operations on the spatial database. These can be used to transform data or undertake analyses. The presence of such tools can be said to be a defining characteristic of a true GIS. If we removed the Manipulation and analysis block from Figure 1.2 the resulting diagram could equally represent computer-aided mapping (CAM) or computer-aided design (CAD) software. Manipulation and analysis tools are what give GIS its unique identity. In more advanced systems, the manipulation and analysis functions will be available directly, and users will also have a programming interface available, to automate repetitive sequences of tasks and to write new analysis functions.

In brief, data manipulation refers to the ability of a GIS to generate new layers from existing ones. This leads to a distinction between *primary* layers, such as the location of roads, rivers or archaeological sites and *secondary* layers, which are derived from these. Primary layers are often digitised from maps, or imported directly from databases. A good example of a secondary layer is the composite soil map created by van Leusen in his study of the South Limburg area of the Netherlands. In creating his spatial database van Leusen quickly realised that his primary soil layer (derived as it was from a modern soil map) contained a number of blank areas corresponding to extensive modern settlement. In an attempt to 'fill in the blanks' he combined this with another primary layer corresponding to a 1920s Geology map to generate a fuller secondary soil type layer (van Leusen 1993:109). Geographic products of primary layers are also possible, for example a secondary layer might be generated from a rivers layer to hold distance bands to rivers, another exercise undertaken by van Leusen (ibid:110). Data manipulation and analysis will be discussed in detail in Chapter 4. One specific primary layer, the Elevation model, has many useful secondary products including terrain slope and aspect. Elevation models will be discussed in detail in Chapter 5.

1.7 WHERE DID GIS COME FROM?

As a number of commentators have pointed out, in its generic-commercial form, GIS is remarkable in that it has received widespread and uncritical disciplinary acceptance despite having a poorly developed archive and virtually no critical history of its own production (Pickles 1995). Although certain aspects of GIS owe their existence to surveillance and mapping initiatives undertaken by NASA and the US military, the origins of modern GIS lie in each of the different areas of computing outlined earlier, with particular emphasis upon computer-aided cartography. Although mainly developed from the 1960s onwards, the idea of using computers to manipulate cartographic data was actually tried as early as 1950, when the Institute of Terrestrial Ecology in Huntingdon, England used the 'Power-Samas Card Calculating System' to map British plant life: processing, sorting, and mapping two thousand species of plant[1]. During the 1960s and 1970s, a variety of computer programs were written to allow the creation of simple geographic maps from digital data. Maps created in this way could, for the first time, be reproduced on a computer screen, plotter or line printer.

Early cartographic programs were capable of jobs such as producing contour plots, generating 3D net plots and interpolating between known points. The Harvard Laboratory for Computer Graphics wrote a number of these early programs (including SYMAP and GRID) and several were used in archaeology, particularly in the United States (several examples are given in Upham 1979). SYMAP was typical of these programs. It required 'a medium-sized computer' and a line printer and was capable of accepting input from punch cards, tape or magnetic disk. These provided the program with a set of co-ordinates defining the boundary of the map area, a second set of co-ordinates relating to the actual data locations, a set of values which corresponded with these points and a set of cards specifying the options to be used. SYMAP was capable of accepting ungridded point data and producing gridded data as output—in other words interpolating from the input points (we discuss interpolation with more contemporary systems in Chapter 9). As with most pioneering computer graphics software, it relied on the capabilities of character printers for output and although it could theoretically produce isoline, choropleth or 'proximal' maps, some imagination was required to interpret the overprinted characters as maps of any kind.

For the very first system that we would recognise as a GIS, we have to look to Canada in the early years of the 1960s. Imagine being faced with the following problem:

"You are part of a team working for the department of natural resources and development for a large country. Among your many duties in managing the resources in your area is to inventory all the available forest and mineral resources, wildlife habitat requirements, and water availability and quality. Beyond just a simple inventory, however, you are also tasked with evaluating how these resources are currently being exploited, which are in short supply and which are readily available for exploitation. In addition, you are expected to predict how the availability and quality of these resources will change in the next 10, 20 or even 100 years.

All of these tasks must be performed so that you or your superiors will be able to develop a plan to manage this resource base so that both the renewable and non-renewable resources remain available in sufficient supply for future generations without seriously damaging the environment. You have to keep in

---

[1] Maurice Wilkes – who also built the first stored program computer – was a consultant on the design of this system, which is now in the collection of the Boston Computer Museum. The author gratefully acknowledges the Boston Computer Museum, Boston, MA for this information.

Figure 1.3 Digitising a paper map using a large format digitising tablet (Rob Hosfield, Archaeological Computing Research Lab at the University of Southampton).

mind, however, that exploitation of resources frequently conflicts with the quality of life for people living in or near the areas to be exploited, because noise, dust, and scenic disturbances may be produced that impact the physical or emotional health of local residents, or devalue their property by creating nearby eyesores. And, of course, you must comply with local, regional, and national legislation to ensure that those using the resource base also comply with the applicable regulations." (DeMers 1997:4-5)

This is clearly a complex, dynamic, multivariate and profoundly spatial problem not dissimilar to the problems facing many cultural resource managers. This was exactly the problem facing the Canadian Department of Forestry and Rural Development at the beginning of the 1960s. In an attempt to develop a flexible and effective means of addressing such problems the Canadian government's Regional Planning Information Systems Division produced the first GIS ever built: the Canadian Geographic Information System (CGIS), implemented in 1964 (Peuquet 1977, Tomlinson 1982).

During the 1970s, many US state and federal agencies wrote similar systems with the result that software for cartography, image processing and GIS advanced considerably during this period. Better output was becoming available from pen plotters and electrostatic printers, input could be obtained from digitising tablets or scanners, and advances in processing speed and interface designs allowed rapid viewing and updating of data.

The motivations of the US agencies and Universities who created these programs were closely related to the massive urban and rural redevelopment schemes of the post-war years. These placed enormous pressures on both surveyors and cartographers to produce and update maps, which were required by planners whose subsequent work immediately rendered such maps obsolete. Essentially, computer automation of cartography was seen as a way to make maps more rapidly and cheaply, sometimes without the need for skilled staff. Storing map data in computer forms also allowed map-makers to generate 'one-off' maps, tailored to particular users' needs, and to make many different kinds of maps without the time a skilled cartographer requires to prepare a new sheet.

From the 1970s onwards there was a gradual move towards the commercial sector. GIS underwent a gradual shift away from being systems designed and written for specific purposes, often by government agencies such as the Canadian Regional Planning Information Systems Division or the US Corps of Engineers, who wrote a raster-based GIS called the Geographic Resources Analysis System (GRASS[2]). One of the earliest commercial companies involved in this was the California-based Environmental Systems Research Institute (ESRI), who began selling a vector-based GIS in the early 1970s. Throughout the 1970s and 1980s more and more commercially supported software products became available until today there are at least a dozen fully functional, commercial GIS systems running on all types of computer platform.

1.8 WHAT DOES IT DO THAT MAKES IT SO ATTRACTIVE TO ARCHAEOLOGISTS?

Given that so much importance has been assigned to the spatial component of the archaeological record, and that considerable methodological development has taken place in its increasingly rapid and accurate collection, it is interesting to note that until very recently the mechanisms used for its presentation, analysis and interpretation had remained largely static. Archaeologists have variously sought to explain spatial organisation through factors such as adaptation to environmental conditions; territorial control; visible patterns in the positions of the heavenly bodies; carefully structured and formalised intent; the physical manifestation of a symbolic-ideological design; the result of a number of quasi-random post-depositional forces; the unplanned, accumulated and embedded product of everyday practice; pure chance; or some combination of all of the above.

The means through which such diverse hypotheses have been explored, however, have been remarkably uniform — the medium of static distribution maps. Whilst the latter are undoubtedly useful heuristic devices in helping to highlight and identify broad spatial trends, distribution maps are often far from satisfactory as mechanisms for exploring complex spatial phenomena. We are rarely interested

---

[2] GRASS remains a free and actively developed GIS for the Unix operating system. It has functionality that puts many commercial systems to shame, albeit with a rather dated interface. There is an active GRASS users list on the internet, which occasionally receives some quite peculiar requests from new contributors.

solely in *where* things are. We often have a wealth of information *about* the artefacts under study—the attribute data we mentioned earlier—and *all* of this information needs to be integrated during the process of interpretation. By simply plotting the spatial dimension on a distribution map we can show the location of things but say very little about what they are or give an indication as to their temporal status. Traditional strategies within distribution mapping for integrating this information have relied upon devices such as the generation of multiple distribution maps, the use of symbols, or both. These solutions are satisfactory until large volumes of information or complex datasets need to be examined. In such cases researchers can rapidly find themselves dealing with hundreds of separate plots or a bewildering array of symbols.

Further problems arise if we want to incorporate information from outside of the artefactual sphere. For example, environmental factors such as the presence of particular types of soil, or cultural and ideological factors such as political boundaries or the optimum alignment needed to best capture the last rays of the setting midsummer sun.

This point can perhaps best be illustrated with an example. During field survey a group of Roman coins is found in the ploughsoil of a field. The coins show little sign of wear and are of a high denomination. Immediately we find ourselves dealing with spatial information (the coins location) and attribute information (the denomination, degree of wear *etc.*). Initially we may want to find within which modern administrative county the find falls, for the purposes of annotating the relevant sites and monuments register. To do this we could manually overlay a map of the coin location with a map of the modern political boundaries. Equally we could simply plot the location directly onto the political map. If we were then curious as to why the coins occurred where they did, we may want to compare the coins' location with that of other contemporaneous coin finds. Our task would be greater but still manageable through studying the relevant monuments register and producing either another manual overlay or adding further information to our original coin location map. We may then want to go further, and extract data from the monuments register to delineate all areas within the immediate environs of contemporary settlement sites (say within 500m) and in close proximity to the communications networks (roads and tracks) to explore the possibility that the coins were lost accidentally rather than the result of deliberate deposition. Although this could be undertaken using manual overlays or distribution maps it would be a time-consuming and complex operation that would need to be repeated if we subsequently decided that perhaps rivers and streams should be included in our category of communications routes as well. We may then want to go even further and identify all of those areas of the landscape, such as springs and ritual sites, which may have attracted ritual deposition. The aim would be to integrate this information with that of the settlements and communication route evidence to create a new map showing the areas most likely to have accumulated losses (whether accidentally or as a ritual act) against which our coin locations could be compared. At this point we find ourselves moving away from the capabilities of paper alone and one step closer to the multi-faceted type of problem facing the Canadian Forestry service.

This is further exasperated if we want to compare the coin locations with this new map. Are the coins significantly clustered within our expected loss/deposition zone or not? We have already discussed how archaeologists in the 1960s and 1970s objected to the uncritical and highly subjective way in which such maps could be interpreted, with the degree of subjectivity directly proportional to the amount of spatial information the eye has to process. If we wanted to examine the statistical significance of any relationship between coin locations and loss zones and identify any possible clustering or grouping within the coin locations, membership of which could be incorporated back into the original attribute database for each coin, we have moved beyond the routine capabilities of distribution mapping and manual overlay.

For archaeologists to make best use of their carefully recovered and recorded information, what is needed is a dynamic and flexible environment within which to integrate, express, analyse and explore the full range of data, both spatial and attribute. Such an environment would ideally permit the vast quantities of collected data to be managed, would enable visual summaries to be generated and would provide a firm platform upon which more sophisticated exploratory and statistical investigations could be launched. It would also be an environment within which you are not restricted in terms of the types of information you could incorporate. Data could relate equally to archaeological artefacts, environmental factors, modern cultural boundaries, perceptual fields *etc.*: in effect, an environment in which to think and explore ideas. GIS has the potential to provide precisely this type of environment of integration and exploration.

1.9 THE DEVELOPMENT OF GIS APPLICATIONS IN ARCHAEOLOGY

Cartographic and spatial analysis software first began to be used for archaeological analyses during the 1970s. Computer graphics and statistics programs, particularly SYMAP, were used to calculate and display trend surfaces calculated from observations of archaeological data at known locations. These software programs were used to implement the statistical approaches being advocated at the time by the New Archaeology movement discussed earlier. Input data generally consisted of the density of some material, such as pottery or bone at a series of locations (*e.g.* Feder 1979), or a measurement such as settlement size or date, which was held to vary over space (*e.g.* Bove 1981). Sometimes digital elevation models were used to display the results of analyses (Arnold III 1979, Kvamme 1983a, Harris 1986). At this time we also see some clear precursors to the types of GIS application we are familiar with today. In his work on early-mid Helladic settlement in the Messenian region of southern Greece, Chadwick generated, combined and explored thematic layers containing archaeological data yielded by field survey and environmental variables such as geomorphology and water supply (Chadwick 1978).

The use of GIS in archaeology began in earnest in the early 1980s, first in the United States (Kvamme 1983a) and then slightly later in the United Kingdom (Harris 1985, Harris 1986) the Netherlands (Wansleeben 1988) and other parts of Europe. In the United States, it was the potential that GIS unlocked for predictive modelling which initially provided the greatest impetus to its widespread adoption (see papers in *e.g.* Judge and Sebastian 1988, Allen *et al.* 1990, Westcott and Brandon 2000).

The attraction of predictive modelling to North American archaeologists may have stemmed from the particular requirements of cultural resource management in some parts of North America and Canada. This often involved management of large geographic areas under dynamic conditions of development or exploitation of which only small parts had been surveyed for archaeology. With large extents of territory essentially unknown to the archaeologists charged with their management, predictive modelling was seen as an ideal aid for the formulation of management plans for National Parks and other wilderness areas. More recently some of these locational modelling methods have been adopted for cultural resource management use within European archaeology, particularly the Netherlands, which is subject to highly extensive and proactive campaigns of redevelopment (Brandt *et al.* 1992, van Leusen 1993). Perhaps because there are not extensive tracts of unsurveyed areas in most European countries, these kinds of models have not proved so popular in the European context. However, the methodology may yet prove to be useful in other parts of the world and for the interpretation of the distributions of some data, such as that derived from fieldwalking (Bintliff 1996, Wheatley 1996a, Gillings and Sbonias 1999, Lock and Daly 1999) or sites discovered through aerial photography. The techniques of GIS-based predictive modelling will be discussed in detail in Chapter 8.

One of the most notable characteristics of the adoption of GIS within archaeology has been the parallel recognition of the importance of the technology by both research archaeologists and by cultural resource managers. Once again the influence of pioneering archaeologists such as OGS Crawford can be detected in the development of methods for keeping inventories of Sites and Monuments. These have foregrounded the importance of spatial information through the tradition of recording sites as location marks on maps, linked to card references of attribute information. Considerable impetus was given to the adoption of GIS by archaeologists involved in national/regional archaeological inventories because existing archaeological records were based on such a system of maps linked by identifiers to record cards[3]. Many archaeologists found that maps of archaeological remains were an intuitive and useful method of representation that was often lost when records were translated into traditional database systems. GIS, which offer a map-based representation of site locations as a primary interface, were therefore a particularly attractive technology. By the beginning of the 1990s, many archaeologists responsible for regional archaeological records were evaluating GIS and at the time of writing cultural resource management organisations as diverse as the Arkansas State Archaeological Department and Hampshire County Council (see Chapter 11) have adopted GIS as the basis for their site inventories.

At the same time, European and American archaeologists whose central interests were in the analysis of archaeological material were also developing a keen interest in the potential of GIS, noting its ability to develop methods of analysis which were previously impossible. Stimulated by a collection of papers published in 1990 as *Interpreting Space* (Allen *et al.* 1990), an international conference on GIS in archaeology was held at Santa Barbara, California in January 1992 (Aldenderfer and Maschner 1996). This was quickly followed by the 1993 Visiting Scholar's conference held at the University of Southern Illinois,

---

[3] In Britain, this dates to the system created by the Ordnance Survey's appointment of OGS Crawford as the first archaeological officer in the 1920s. The Ordnance Survey's archaeological record later became the basis of both County Sites and Monuments Records and the National Archaeological Record of the Royal Commissions for Historic Monuments.

Carbondale in 1993 (Maschner 1996a). These meetings served to stimulate academic interest and widen archaeological exposure to the technology.

Until the early 1990s the majority of interest in GIS applications in archaeology had been from North American archaeologists. Trevor Harris introduced GIS to the British archaeological literature in 1986 (Harris 1986) but subsequently took a position in the United States. As a result it was not until Gaffney and Stancic demonstrated the value of GIS in the interpretation of regional survey data (Gaffney and Stancic 1991, 1992) that a significant number of European archaeologists began to take serious interest. Following the Santa Barbara conference, which was attended by several European archaeologists, an international conference was held at Ravello, Italy in 1993 specifically to showcase work taking place in Europe (Lock and Stancic 1995).

What all of these conferences clearly showed was the huge range and variation of applications of GIS to archaeological problems. As well as making much existing work on predictive modelling available to a far wider audience, the conferences and their subsequent publications demonstrated how GIS could revolutionise existing areas of study such as the interpretation of regional survey data and site catchment analysis. Perhaps more importantly, they also showed how the technology could open up new avenues of research as diverse as visibility analysis (Ruggles *et al.* 1993, Wheatley 1995a, van Leusen 1999, Wheatley and Gillings 2000) and flood dynamics (Gillings 1995, 1997).

Since the early 1990s, there has been a continuing growth in the number of archaeological projects that make use of GIS, to the extent that it would now be difficult to conceive of a regional survey project or spatial modelling exercise that was not based around spatial technology. At the same time, some debate about the relationship between GIS-based archaeological analyses and wider issues of archaeological theory have also surfaced (Zubrow 1990b, Wheatley 1993, Harris and Lock 1995, Wheatley 2000, Wise 2000). These have often centred on the relationship between predictive modelling and environmental determinism (Gaffney and van Leusen 1995, Kvamme 1997, Wheatley 1998) with some archaeologists arguing that unthinking use of GIS can lead unwittingly to interpretations of culture that overemphasise the impact of environment factors as the prime mover for cultural activity.

1.10 CONCLUSION

In this chapter, we have looked at how archaeologists have thought about and conceptualised space and sought to depict and utilise spatial information. We have made it clear that thinking about space did not start with the advent of spatial technologies in the discipline, but instead has a long history that is intimately related to wider developments within archaeological thought. We have also offered a broad definition of GIS, and have looked in outline at its relationship to archaeology.

We have also given an overview of the debates about the theoretical implications of GIS-based archaeology. It is our view that these should be seen as a healthy sign of reflectivity and self-criticism: regardless of theoretical differences, the utility of GIS and the variety of new avenues of research which it offers to archaeology continues to broaden its appeal within both archaeological research and management.

Having examined what GIS is, where it came from and what it does that makes it so attractive to archaeologists, our next task is to look in more detail at how GIS actually works.

CHAPTER TWO

The spatial database

"When considering any space—a room, a landscape, or a continent—we may adopt several fundamentally different ways to describe what is going on in that subset of the earth's surface." (Burrough and McDonnell 1998: 20)

So far, we have discussed GIS in a very broad sense. From here onwards, we begin to look in more detail at how precisely the GIS works. As mentioned in Chapter 1, central to any archaeological application of GIS is the *spatial database*. We use the term to define the entirety of information we have held in the GIS for a certain study area. Any given spatial database must provide for the storage and manipulation of four aspects of its data:

- A record of the position, in geographic space (the *locational component*) that determines where something is and what form it takes;
- A record of the logical relationships between different geographic objects (the *topological component*);
- A record of the characteristics of things (the *attribute component*) that determines what geographic objects represent, and what properties they have;
- Thorough documentation of the contents of the overall database (the *metadata component*).

2.1 HOW DOES A SPATIAL DATABASE DIFFER FROM A TRADITIONAL DATABASE?

Traditional (non-spatial) databases are concerned only with the attributes of objects. As a result they make no explicit distinction between the location of an object (*e.g.* an archaeological site) and its other attributes (date range, finds, structural evidence *etc.*). An example would be the common practice of storing a grid reference for each site in an inventory. This distinction is a crucial one and can be illustrated with a hypothetical example. Consider the feature complex shown in Figure 2.1. Here, we have an Iron Age Hillfort comprising two concentric ramparts. Inside the outer rampart, but overlain by the inner is an earlier Neolithic bank barrow. Abutting the Hillfort are the remains of a prehistoric field system. Although fictional, in the UK this is not a particularly artificial or unusual configuration of archaeological remains.

In a traditional monuments inventory or register based on a conventional database, there would be individual entries for the Hillfort, the barrow and the field system, in each case giving some details about the physical remains, the legal status, bibliographic details, history of investigation and so on. At a minimum, the database would also give a map reference for each monument. It is not uncommon for features such as the field system to have several different entries, referring to a number of discrete locations spaced at regular intervals throughout the overall

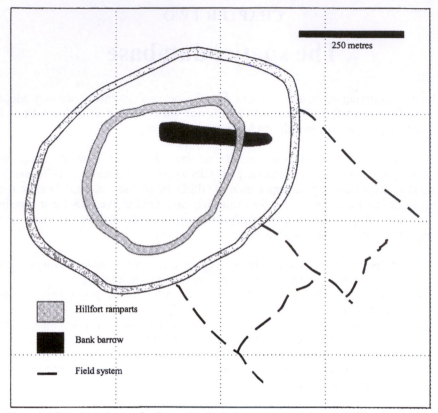

Figure 2.1 Hypothetical example of an archaeological complex consisting of a Hillfort, bank barrow and adjoining field system.

complex. Now consider the following enquiries, which might not unreasonably be made by an archaeologist, resource manager or planner.

- *What is the extent, in hectares, of the hillfort?*
- *Is the bank barrow wholly enclosed by the hillfort?*
- *How do the orientations of barrow and field system compare?*
- *What is the relationship between the field system and the hillfort?*
- *How many other monuments are partly or wholly within 1km of the barrow?*
- *How many other monuments are visible from the hillfort?*

Whatever the query facilities available within the monuments register, it is clear that without reference to maps or more detailed textual descriptions of the sites no traditional database would be able to answer these questions. To answer the first requires a geographic representation of the form of the site—the spatial component—while the remainder require that the database has some representation of the logical relationships between the geographic forms of the features themselves or information derived from them (*e.g.* they are in view of the zone of land within 1km of the Hillfort).

The situation becomes even more complex when an archive or inventory is required to hold aerial photographs of the sites or to retain a record of fieldwalking data or geophysical survey. Each of these kinds of data requires an approach to storage that is explicitly based on geographic location.

In the following sections we will concentrate on the first two aspects of the spatial database highlighted in the introduction: the locational component (how the GIS records and represents the positions of features in space); and the topological component (how it encodes and interprets the geometrical relationships between features). Attribute and metadata will be discussed in Chapter 3.

2.2 THEMATIC MAPPING AND GEOREFERENCING

Let us begin by looking at how the spatial database organises and manages spatial information. Rather than storing it in the form of a traditional map, GIS store spatial information *thematically*. This can best be illustrated with reference to a typical map sheet. Any given map contains a wealth of information relating to a host of specific themes. These might be: topography (in the form of contours); hydrology (rivers and streams); communications (roads, tracks and pathways); landuse (woods, houses, industrial areas); boundaries (*e.g.* political or administrative zones) not to mention archaeology (the point locations of archaeological sites). Although we think of maps as a highly familiar way of representing the spatial arrangement of the world, closer perusal reveals the everyday map to be an abstracted and deceptively complex entity encoding an enormous breadth of information. Returning to the GIS, an important point to realise is that GIS systems *do not* conceptualise, store and manage spatial information in such a complex and holistic form. As a result, although it is often convenient to talk of the discrete collections of spatial information held within the GIS as 'GIS-maps' it is formally incorrect.

Instead, GIS rely upon the concept of 'thematic mapping'. Instead of storing a single, complex, multiply-themed map sheet, GIS store and manage a collection of individual sheets, each themed to a particular facet of the region under study. The precise terminology used for these discrete slices of thematic data varies, with some systems using the term *layer* and others *theme*, *coverage* or *image*. To avoid confusion the term *layer* will be used exclusively in this book.

In the example of the hypothetical map mentioned earlier, we would no longer have a single map but instead a group of themed layers: topography (holding just the contours); hydrology (the network of rivers and streams); communications (roads and paths); landuse (the areas of woods, houses and industrial activity); boundaries (the administrative zones) and archaeology (the point locations of archaeological sites). Some of these layers are depicted in Figure 2.2.

The important point to emphasise is that a spatial database and a traditional map sheet can contain the same information, but are structured in a very different way: a set of specific thematic layers as opposed to a complex totality. Let us look at some archaeological examples. In their pioneering work on the Dalmatian island of Hvar, Gaffney and Stancic constructed their spatial database using five basic layers of information: topography; soils; lithology; microclimate and archaeological site locations (Gaffney and Stancic 1991:36). In a recent assessment of archaeological resource sensitivity in Pennsylvania and West Virginia, Duncan and Beckman based their spatial database around the following layers: Prehistoric site location; historically documented Indian trails; roads and disturbance factors; hydrology; soils; and topography (Duncan and Beckman 2000, see Table 2.1).

The use of thematic layers makes a lot of conceptual sense, particularly in the management and organisation of data. For the concept to work we have to be able to combine and overlay the individual layers in our database whilst maintaining the original spatial relationships held between the component features. For example, if on the original map sheet an archaeological site fell 23m from a river and sat exactly on the 92.5m contour, then if we were to ask the GIS to overlay the archaeology, hydrology and topography layers in the spatial database, the same relationship should hold. To ensure that this is so, at the heart of the spatial database is a mechanism whereby every location in each data layer can be matched to a location on the earth's surface, and hence to information about the same location in all the other data layers. This makes it possible for the user of the GIS to identify the absolute location of, for example, a pottery density held in a fieldwalking theme, the limits of a protected archaeological site, a river in a hydrology layer and any other archaeological or non-archaeological information within the overall spatial database.

This registering of the spatial locations of features in the individual thematic layers to the surface of the earth relies upon a concept we term *georeferencing*. The term *georeferenced* refers to data that have been located in geographic space, their positions being defined by a specified co-ordinate system.

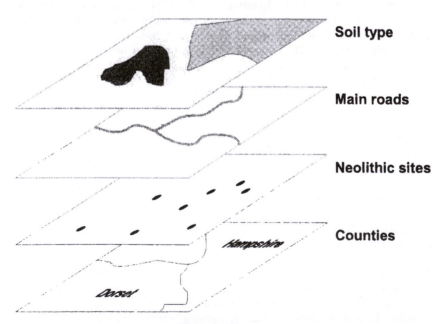

Soil type

Main roads

Neolithic sites

Counties

Figure 2.2 Thematic layers.

Table 2.1 Examples of primary thematic layers (after Gaffney and Stancic 1991:36; Duncan and Beckman 2000:36).

Layers	Gaffney and Stancic	Duncan and Beckman
Layer 1	Topography	Prehistoric site locations
Layer 2	Soils	Historically documented Indian trails
Layer 3	Lithology	Roads and disturbance factors
Layer 4	Microclimate	Hydrology
Layer 5	Archaeological site-locations	Soils
Layer 6		Topography

This can be illustrated with reference to the typical process of archaeological survey. Commonly a baseline or site grid is established and a series of measurements are taken with respect to it. This could be to record the locations of finds in an excavation trench or capture the shape of an earthwork feature. Once recorded, the measurements are then used to produce a scaled drawing of the features ready for interpretation and publication. Although the measurements and subsequent drawing are spatially accurate with respect to the trench location or morphology of the features, we have no idea where they are in relation to the wider world. The only spatial reference we have has been internal to the survey itself — the survey grid or baseline. We could not compare the location, size and extent of the earthwork to any other features in the area, as we have no way of relating their relative positions. Such surveys are called *floating* or *divorced* surveys for this very reason. In GIS-based studies, georeferencing is a vital part of any spatial database. This is because it is through georeferencing that data encoded on separate thematic layers can be combined and analysed. It is also vital for the interpretation of the results of GIS-based analyses because it allows the products of GIS to be related to objects and locations in the real world.

So far we have established that the term georeferencing refers to the location of a thematic layer in space, as defined by a known co-ordinate referencing system. Two types of co-ordinate system are currently in general use. The eldest of these is the geographical co-ordinate system, based upon measurements of latitude and longitude. This is the primary locational reference system for the earth, and measures the position of an object on the earth's surface by determining latitude (the North-South angular distance from the equator) and longitude (the East-West angular distance from the prime meridian—a line running between the poles and passing through Greenwich observatory near London). In this system, all points on the earth falling at the same latitude are called a parallel, whilst all points falling at the same longitude are said to be on a meridian.

The other type of co-ordinate system in widespread use is the rectangular or planar co-ordinate system. Planar co-ordinates are used to reference locations not upon the curving surface of the earth but instead upon flat maps. Modern plane co-ordinate systems have evolved from the concept of Cartesian co-ordinates, a mathematical construct defined by an origin and a unit of distance. In a plane, two axes are established running through the origin and spaced so that they run perpendicular to each other. These are subdivided into units of the specified distance from the origin. Good examples are the international UTM grid system and the Ordnance Survey grid covering the United Kingdom (Gillings *et al.* 1998:14-15). If all of the spatial features of interest, recorded on all of our component thematic layers are referenced using the same co-ordinate system, then they can be overlain and combined maintaining the accuracy and integrity of the spatial relationships encoded upon the source map.

The establishment of a co-ordinate system in turn relies upon a process of what is termed *projection*. Projection refers to the mathematical transformation that is used to translate the irregular curved form of the earth's surface to the two-dimensional co-ordinate system of a map. Such a translation cannot be undertaken without some compromise in properties such as area, shape, distance and compass bearing. Needless to say, the precise projection used has an important influence upon the properties of the map derived from that process. As a result, understanding which projection has been used in the creation of a map is an essential first step in incorporating it into a spatial database.

For very small study areas, it is sometimes acceptable to ignore projection, and to assume that the region of interest does correspond to a flat two-dimensional surface. Most geophysical survey software takes this approach because surveys are rarely larger than a few hectares. However, if the study region is larger than a few kilometres, or if information is to be included from maps which have been constructed with different projections, then to be able to georeference the thematic layers held within the overall spatial database, the GIS needs to understand the projection used for each layer in order to avoid inaccuracies.

2.3 PROJECTION SYSTEMS

The mathematics of projection are beyond the scope of this text, but it is worth considering briefly the major types (or families) of projection in order to understand the essence of the process. Map projections consist of two main stages, which we will examine in turn.

Firstly, the surface of the earth is estimated through the use of a geometric description called an ellipsoid. This is often referred to as a spheroid, though such a designation is not always formally correct (Gillings *et al.* 1998:13). Unlike a circle, which boasts a constant distance from its centre to any point on its edge, in an ellipsoid the distance is not constant. As a result, whereas we define circles on the basis of their radius and diameter, ellipsoids are defined in terms of their equatorial radius (what is formally termed the semimajor axis of the ellipse) and by another parameter, such as the flattening, reciprocal flattening or eccentricity. Another common term applied to geometric figures used to estimate the earth's form is *geoid*. Most of the ellipsoids that have been used to generate maps have names such as the '1849 Airy' ellipsoid used by the Ordnance Survey of Great Britain or the 'Clarke 1866' ellipsoid used with the UTM system in the United States. Fortunately, most GIS provide a list of known ellipsoids that can be selected by the user when data is entered.

Having approximated the shape of the earth, the second stage in the process is to project the surface of the ellipsoid onto a flat surface. There are a huge and varied number of methods for undertaking this process of projection. As intimated

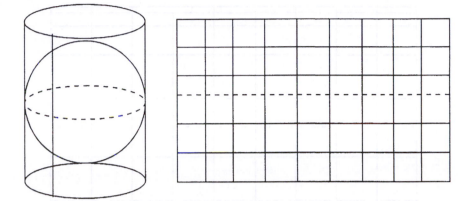

Figure 2.3 Cylindrical projection and graticule.

earlier, each method produces a map with different properties. For example in a *cylindrical projection*, the flat surface is wrapped around the ellipsoid to form a cylinder. An imaginary light source is then placed at the centre of the ellipsoid and used to project the lines of latitude and longitude on the surface of the enclosing cylinder. The lines of latitude (parallels) of the selected spheroid are then drawn as simple straight, parallel lines.

As can be seen in Figure 2.3, the lengths of the lines of longitude (parallels) .are progressively exaggerated towards the poles. To maintain the right-angled intersections of the lines of latitude and longitude, the lines of longitude (meridians) are also drawn as parallel lines which means that the meridians never meet at the poles, as they do in reality, so that the polar points of real space become lines the length of the equatorial diameter of the ellipsoid in projection space. The cylindrical projection thus maintains the correct length of the meridians at the expense of areas close to the poles that become greatly exaggerated in an east-west direction. A *transverse cylindrical projection* is created in the same manner, but the cylinder is rotated with respect to the parallels and is then defined by the meridian at which the cylinder touches the ellipsoid rather than the parallel. Reducing the distance between the lines of latitude as they move away from the poles can offset the exaggeration of the areas at the polar regions. This is called a *cylindrical equal-area* projection (Figure 2.4). However, this projection distorts the shapes of regions very badly, and cannot represent the poles at all because they are projected into infinitely small areas.

One other cylindrical projection is worth mentioning. This is the *Mercator* projection, which exaggerates the distance between meridians by the same amount as the parallels are exaggerated in order to obtain what is termed an *orthomorphic* and *conformal* projection. The term orthomorphic describes the fact that shape is preserved so that squares in the projection plane are truly squares. Conformal means that the parallels and meridians in the projection plane always intersect at right angles. The poles cannot be shown because they are infinitely distant from the equator, and the projection exaggerates the apparent areas of northerly areas far more than either of the previous projections—at 60 degrees latitude this exaggeration is 4 times, at 70 degrees 9 times and at 80 degrees it becomes an exaggeration of 32 times the same area if it were shown at the equator. A

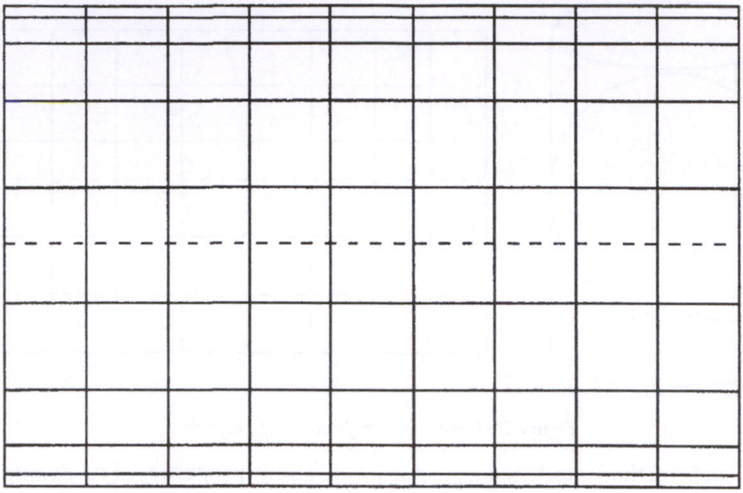

Figure 2.4 Cylindrical, equal-area projection graticule.

transverse Mercator projection is similar except that it is based on the transverse cylindrical projection. This means that instead of the north-south exaggeration of the traditional Mercator projection, the distance of areas to the east and west of the defining meridian are exaggerated in the projection plane.

Cylindrical projections serve to illustrate some of the distortions that are introduced into the representation of space as a result of the projection process, but practitioners should be aware that there are a great many other forms of projection that are not based on the cylinder. Conical projections (such as the *Lambert Conformal Conic*) are based on a cone which intersects the spheroid at one or more locations. Where the point of the cone is above one of the poles, then it intersects at one or more parallels (termed standard parallels). As can be seen in Figure 2.5, conical projections depict the parallels as curved lines, which decrease in size as they approach the poles. The meridians are shown as straight lines that meet at the poles. In other projections, the meridians are curved. A cylindrical projection where all of the parallels are correctly drawn and divided, for example, results in an

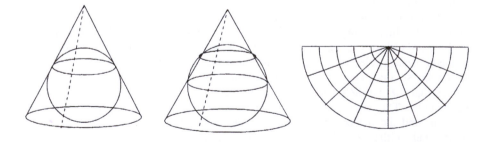

Figure 2.5 Conical projection with one (left) and two standard parallels (middle) and the general form of the resulting projection (right).

equal area projection plane with curved meridians called *Bonne's Projection*. Two entirely separate families of projections called *two-world equal area projections* and *zenithal projections* are also occasionally used for particular purposes.

2.4 FURTHER COMPLICATIONS

To further complicate matters, it is not uncommon for a basic projection to be modified through the use of correction factors. For example, in transverse Mercator projections it is not uncommon for a scaling factor to be incorporated into the east-west axis, which serves to correct for the east-west distortion inherent in the projection itself. It is also common to use a false origin: in other words to arbitrarily define some x, y location on the projection plane to act as the origin point 0,0. If the true origin of the projection is far from the actual mapped area, the false origin can be placed on the positive x and y axes. This makes the axis numbers far shorter and easier to use. Where the origin of the projection occurs within the mapped area, it is not uncommon for a false origin to be defined on the negative side of the x and y projection plane, to ensure that all of the co-ordinates quoted are positive. False origins are used by both the UK Ordnance survey and UTM co-ordinate systems.

The most important point to understand is that all projections result in spatial distortion and different projection systems distort in different ways. If we are aiming to integrate spatial data from a variety of sources into a single GIS-based spatial database, the GIS must be aware of the projections used in their initial creation. Different parameters will need to be given to the GIS depending upon exactly which projection has been used.

To recap: the GIS stores, manages and manipulates spatial data in the form of thematic layers. For the thematic-layer structure to function, the component themes must be georeferenced, ideally to the same co-ordinate system. To be able to integrate spatial data from a range of map-based sources within a single spatial database care must be taken to ensure that they either derive from the same projection or that the GIS holds and understands the relevant georeferencing and projection information about each individual data layer. This is in order to determine how different primary layers which might be represented at different scales or with different projections, relate together in space.

It is also necessary in order for the GIS to be able to undertake calculations of spatial parameters such as distance and area. In some instances it is possible to assume that the layers can be manipulated by using simple Euclidean geometry, either ignoring effects of the curvature of the earth as acceptable errors, or accepting that the selected projection plane has appropriate properties. However, where the locations of entities are recorded in different co-ordinate systems it is necessary for the system to store both the co-ordinates of the entities and the projection through which they are defined. If this is not undertaken, and the characteristics of the projection plane used are not incorporated into the information system, errors in distance and area calculations are introduced.

We can think of the projection information related to a given layer as being critical for us to be able to effectively integrate and use it. Another term for describing such essential data-about-data is *metadata*. We will be looking in more detail at metadata in Chapter 3.

PROJECTION EXAMPLE: THE UK NATIONAL GRID

To understand how projection relates to the use of maps in a GIS, we really need to consider a familiar mapped space. For the authors, the most familiar maps are those made of the United Kingdom by the Ordnance Survey of Great Britain. Superficially, these seem straightforward enough: we can simply digitise them (leaving aside the issue of copyright for a moment) and assume that the shapes and areas will be those of a flat plane. However, the reality is a little more complex than that.

In the United Kingdom, a *transverse Mercator* projection has been used as the basis for all recent maps. Transverse Mercator projections are particularly suitable for areas with a north-south extent as the north-south distortion of areas is removed in favour of an east-west distortion. The UK's projection uses the 1849 *Airy spheroid*, which estimates the earth as an ellipsoid with equatorial radius of 6377563.396 metres and a polar radius of 6356256.910 metres and the projection is made from the 2-degree (west) meridian.

The national grid is defined in this projected space by defining a false origin at crossing of the 2 degree (west) meridian and the 49 degree (north) parallel, in which x is 400000 metres and y is -100000 metres. A scaling factor of 0.996 is also used.

For small areas, any errors of shape and area will be minimal, but if we are concerned with a significant area of or even the whole of the UK then we should represent all of our data in a geodetic (longitude, latitude) system and then instruct the GIS how to translate between them. As you may now appreciate, this is not necessarily straightforward.

2.5 SPATIAL DATA MODELS AND DATA STRUCTURES

The precise way in which a given GIS conceptualises, stores and manipulates spatial information is referred to as its *spatial data model*. Rather than a single, generic GIS, there are currently two types in common archaeological usage, which differ in the precise spatial data model they implement. These two basic types are termed *Vector*-GIS and *Raster*-GIS. The essential difference is that vector data is a formal *description* of something in the real world, usually in the form of geometric shapes. Raster data on the other hand comprises a number of *samples* of something, usually taken at regularly spaced intervals.

Raster systems (Figure 2.6 right) store information as a rectangular matrix of cells, each of which contains a measurement that relates to one geographic location. Raster images thus constitute a 'sampling' approach to the representation of spatial information. The more samples that are taken, the closer the representation will be to the original. One of the most widespread forms of raster data used in GIS is remotely-sensed imagery, from earth orbiting satellites such as SPOT and Landsat Thematic Mapper (see Chapter 3). In this case, each of the raster cells in the resulting matrix is a measurement of the amount of

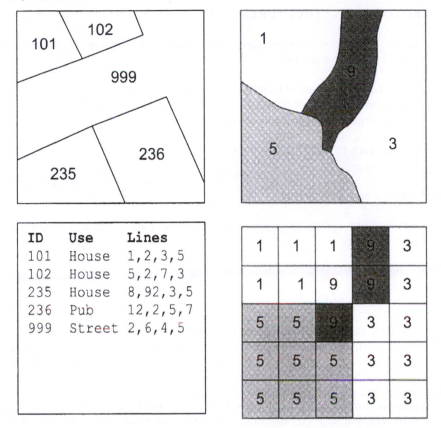

Figure 2.6 Vector (left) and raster (right) representations of an area map.

electromagnetic radiation in a particular waveband that is reflected from a location to the sensor.

In contrast, in a vector data file (Figure 2.6 left) the data is represented in terms of geometric objects or 'primitives'. These primitives are defined in terms of their locations and properties within a given co-ordinate system, or model space. Vector systems do not so much sample from an original as describe it. The most common origin of vector data is from digitised maps, where the various features inscribed upon the surface of the paper map are converted into co-ordinates, strings of co-ordinates and area structures, usually through the agency of a digitising tablet.

It should be appreciated that within each of these basic data models there are ways in which data models may be implemented, often closely tied to specific software packages. As a result of this diversity, in the following discussion the aim has been to concentrate more on generic properties and features of the models themselves rather than specific details of individual data structures. Let us begin by looking in detail at the vector model.

2.6 VECTOR DATA STRUCTURES

In a vector data structure each geographical entity in a given layer can be represented by a combination of:

- A geometric primitive (fundamental mappable object);
- Attribute data which defines the characteristics of the object;
- Topological data (discussed further below).

Let us look at each of these features individually.

Fundamental mappable objects

Geographical data are usually represented by three basic geometric primitives (Burrough 1986:13):

- point
- line (arc)
- area (polygon)

The first, and simplest way of representing a feature is as a discrete *x, y* co-ordinate location, in effect a *point*. The second method of representation builds upon this, representing features as series of *x* and *y* co-ordinates, which are ordered, so as to define a *line*, often referred to as an *arc*. The third method of representation builds upon this, representing features as a series of lines which join to enclose a discrete *area*, normally referred to as a *polygon*. In each case the basic building blocks are discrete *x, y* co-ordinates. In the case of points they are treated in isolation; with lines they are ordered to form a sequence; and in the case of areas ordered to form a set of lines that enclose a discrete area. These basic geometric primitives can be termed *fundamental mappable objects* and are depicted in Figure 2.7.

In the vector-GIS the thematic layers in the overall spatial database comprise sets of georeferenced points, lines and areas. As well as storing the co-ordinate information relating to each discrete feature in a given layer, the vector-GIS assigns each feature a unique code or identification number. This code number is critical as it not only enables the GIS to keep track of the various features within each layer, but also provides a vital link that can be used to relate attribute information to the features.

To illustrate how these fundamental mappable units are applied in practice, a linear earthwork bank might be represented as a line with the attribute 'Earthwork bank' and information about its date, period and legal status. An isolated lithic find-spot may be represented as a point object, whose attributes include the label 'Obsidian', some information about its context, its dimensions and perhaps techno-functional information; while an area of limestone geology could be represented as an area entity with the label 'Limestone' and information about the calcium content or geological period.

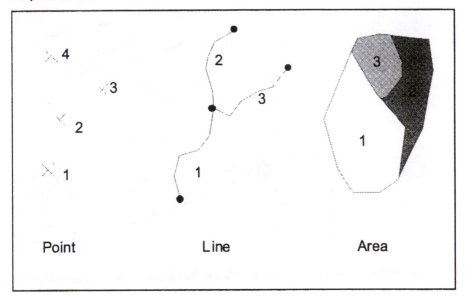

Figure 2.7 Examples of point, line, area.

Spatial dimension of primitives

The basic geographic elements discussed above are often referred to in terms of their dimensionality. A point entity is regarded as having no spatial extent, and is thus referred to as a 'zero-dimensional' spatial entity. Lines connecting two points possess length, but not extent and are therefore referred to as 'one-dimensional' entities. 'Two-dimensional' spatial objects are those that have both length and width—in other words areas. The spatial dimension of an entity is important because it has a direct effect on the logical relationships that are possible between geometric entities.

Imagine a spread of points. Because a point is zero-dimensional, it cannot contain or enclose other entities. Now add a line. Because lines are considered to have only length, but no width—one-dimensional—the points entities must either be on one side of the line or the other.

Intuitively it may be expected that surfaces, such as elevation models, would be three-dimensional. However, surfaces in most GIS systems are stored with the third dimension recorded as an *attribute* of a two-dimensional entity. True 3D representation of objects requires surfaces and solids to be represented through an extension of the *spatial component* of the data model into three dimensions, not merely the use of an attribute as a z co-ordinate. To separate the typical GIS-derived surface representations from true 3D models (we discuss fully 3D GIS in Chapter 12), they are sometimes referred to as 2½-dimensional entities.

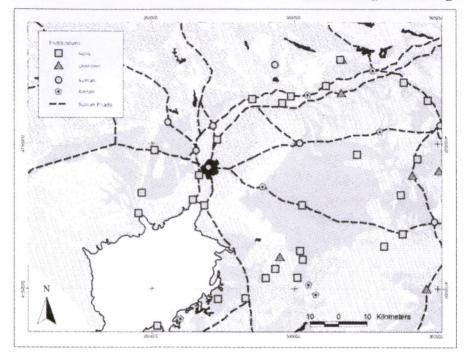

Figure 2.8 Fortified Roman and Iron Age centres in the lower Guadalquivir Valley, Andalucia. This single GIS-produced map combines area, network and point data themes. Point attribute data (nature of the fortifications) is used to control the shapes of the sites and the polygon attribute data (soil classification) to control the type of shading.

2.7 AN EXAMPLE OF A 'SIMPLE' VECTOR STRUCTURE

The simplest method of explaining how vector GIS data files are structured is to work through an example. Rather than adopt a particular proprietary format, however, the description that follows is of a slightly simplified and fictional format. Its structure actually owes something to several real formats used by commercial GIS.

To describe a geographic theme, then, we will need sections of our data file to describe *point, line* and *area* features and we will also need sections for their *attributes*. Before that, however, we will start with a simple format for describing the general properties of our data theme in terms of a *documentation section*.

Documentation section

In our format, this consists of two columns, separated by a comma. The left column identifies the data, which is given in the right hand column. A typical documentation section might read:

```
DOCUMENTATION
    title,     Long Barrows
    system,    plane
    units,     metres
    min. X,    400000.0000000
    max. X,    420000.0000000
    min. Y,    160000.0000000
    max. Y,    180000.0000000
```

The *title* field contains the title of the whole data theme, *system* and *units* indicate the type of co-ordinate system and the unit of measurement for the co-ordinates and the next four fields define the maximum and minimum co-ordinates of the data—in other words they define a rectangular region within which all of the objects will be found.

This is, admittedly, a very minimal documentation section. Most systems will also keep a record of the original scale of the data, the source, the projection system used, some estimate of its accuracy and many other facts about the data theme.

Data section for points

Point data is straightforward to store in a spatial database. Point data normally consists of a series of co-ordinate pairs, each associated with an identifier. In our simple vector data structure, the data section for a points file might follow this format:

```
POINTS
identifier, x1, y1
...
identifier_n, xn, yn
end
```

The POINTS line tells the GIS that each of the lines that follows reference a single point entity. This is followed by any number of identifiers preceding the relevant co-ordinate pair. The identifier is a number which uniquely identifies each of the points, and to which any attribute data can be linked (see below). For example, if we were recording the locations of small-finds in an excavation trench, a sensible identifier would be the unique small finds numbers assigned to each artefact as it is excavated. An identifier string 'end' is used to indicate the end of the section.

A complete data section for the four long barrows shown in Figure 2.9 might be as follows:

```
POINTS
1, 103.97,    98.43
2, 143.75,    103.79
3, 110.85,    93.63
4, 114.17,    89.04
end
```

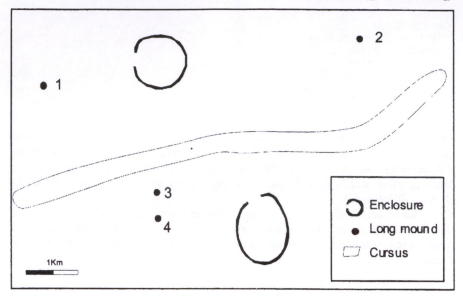

Figure 2.9 A hypothetical landscape.

This contains the codes 1,2,3 and 4 to identify the long barrows and the co-ordinate locations of each of the barrows.

Data section for lines

The additional complication presented by line files (e.g. the cursus shown in Figure 2.9) is that they can contain any number of points. To allow for this, the lines section of our data might require that the identifier be followed by the number of points in the line, and then by that number of co-ordinate pairs like so:

```
LINES
identifier, n
    x1, y1,
    . . .
    xn yn
identifier2, n2
    . . .
end
```

As for points, we can allow an identifier of 'end' to indicate the end of the lines section. An area section (*e.g.* for the enclosure depicted in Figure 2.9) can be identical to the line section except that the identifier will be AREA and the first and last co-ordinates of the line will be the same so that the line segments close to form an area or closed polygon (in other words $x1,y1 = xn, yn$).

Attribute values files

On a traditional paper map, the attributes of the geographical objects are either written as text next to the objects, or defined by a legend. A legend provides a list of attributes together with the symbols, line-types, hatching or colours which are used to identify them on the map. In the spatial database they are more commonly stored as simple ASCII attribute descriptor files (as shown below) or tables in a DBMS. In practice, the identifiers assigned to each object can be used to link the spatial entities with any number of attributes. Our system will allow attributes to be defined in an attribute section, which associates attribute codes with attributes as follows:

```
ATTRIBUTE
name,   type
code,   attribute
code,   attribute
code,   attribute
end
```

The section begins with the name of the attribute and its data type (*e.g.* character, decimal, or integer) and then contains one line for each identifier for which an attribute is recorded. A simple attribute file containing a single piece of information *about* each of the long barrows in the point file described above, might serve to link the identifier code for the long barrows with a number indicating their length in metres:

```
ATTRIBUTE
length, decimal
1,   39.5
2,   24.3
3,   66.8
4,   86.7
end
```

Exactly the same process can be used for assigning attributes to line and area entities. Additionally, there can be several attribute sections in the file, allowing the spatial entities to have any number of different attributes associated with them—for example, different attributes sections might exist to indicate the name, orientation and type of the long barrows. A complete data file defining the four barrows, each with four attributes is listed in Figure 2.10.

Topological information

When we use a map we are often explicitly interested in the logical geometrical relationships between objects. For example, if we were trying to re-locate a ploughsoil scatter on the basis of a shaded area on a 1:10,000 scale map sheet, we would look for the specific field or block of land the site was located in, and perhaps the nearest recognisable building or intersection of field boundaries to orient ourselves within the field.

```
DOCUMENTATION
    title,    Long Barrows
    system,   plane
    units,    m
    min. X,   400000.0000000
    max. X,   420000.0000000
    min. Y,   160000.0000000
    max. Y,   180000.0000000
POINTS
    1, 103.97,   98.43
    2, 143.75,  103.79
    3, 110.85,   93.63
    4, 114.17,   89.04
    end
ATTRIBUTE
    length, decimal
    1,   39.5
    2,   24.3
    3,   66.8
    4,   86.7
    end
ATTRIBUTE
    name, char
    1,   "Adams Grave"
    3,   "Beckhampton Road"
    end
ATTRIBUTE
    orientation, integer
    1,   1
    2,   4
    3,   3
    4,   1
    end
END
```

Figure 2.10 A complete example of a simple vector data structure, defining four point entities, each of which might have up to three attributes (length, name and orientation).

Encoding topology

Topology therefore refers to the logical geometry of a data theme (we discussed this at the start of this chapter, but see Figure 2.11 for a reminder of why topology is a familiar concept to most archaeologists). It comprises the connections and relationships between objects. In the GIS it has to be explicitly calculated and recorded rather than inferred from their spatial location. Let us return to our geometric primitives: points, lines and areas. As points effectively have 0-dimensions they do not really pose any topological problems. However, with lines a number of important potential topological relationships can be identified.

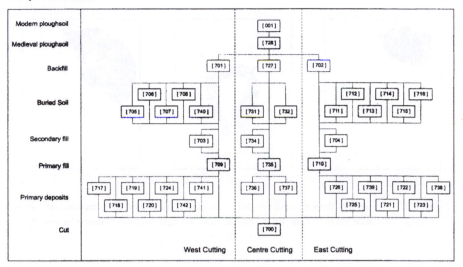

Figure 2.11 A Harris matrix as an example of topology. The true spatial locations of features are irrelevant, what is important are their absolute stratigraphic relationships.

Topology of line data

Consider a simple road network. The locational component of this can be represented as a series of lines. Attributes can be recorded for each of the roads, for example the name of the road as shown in Figure 2.12 (left).

However, this in itself is insufficient to record the full complexity of the road network. If we are interested in the possibilities of transport within this system, we need to know whether we can drive from the A31 onto the M4 for example. To

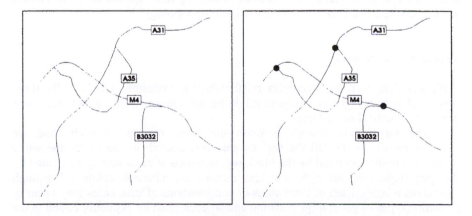

Figure 2.12 Roads represented as line entities with attributes. Right: introducing nodes (black dots) makes the relationships between roads clear.

Figure 2.13 Three regions showing some possible topological relationships. Left, unconnected; centre, regions contained island fashion within other regions; and right, regions connected by shared boundaries.

understand this, we need to record where the roads meet as opposed to where they cross each other with bridges. This sort of information about the relationships between objects is part of the topology of the theme. If we simply store the features (in this case individual roads) as little more than ordered lists of co-ordinates (as we did in our simple vector data structure), we have no way of recording whether and where lines connect, cross or simply terminate without a connection.

In the case of the road network it is necessary to introduce another spatial object into the model to record the way in which the roads relate to one another. We can refer to this as a node. A node represents the point at which an arc begins or ends, and records the other arcs to which it is connected. We will be discussing nodes in a little more detail shortly. In Figure 2.12 (right) a series of nodes have been included corresponding to road junctions, and we can now infer that to get from the A31 to the M4 we must use the A35. Note that our roads can be represented by several arcs—in our example the M4 is composed of three separate arcs—and that lines which cross must be broken up into separate arcs, with nodes at the crossings, if we are to use nodes to build up topology.

Topology of area data

With area data even more complex relationships are possible. Consider the three regions shown in Figure 2.13 (these might be soil classes within a site catchment zone, or geological substrates).

The visual relationships between the three regions in each case are straightforward: on the left the regions are unconnected; in the centre the white region is wholly contained by the black region which is itself wholly contained by the grey region; on the right the three regions are adjacent, connected through shared boundaries. Each of these cases or combinations of such cases may occur in particular types of geographic situation and so each must be explicitly coded in the spatial database.

Problems with simple data structures

To enable the vector-GIS to effectively 'understand' the geometrical relationships between the features in the spatial database it has to establish topology for them. This information has to be stored with each data layer or calculated 'on the fly' as and when needed. It should be noted that there are few standards as to how this topological information should be stored, with different commercial vector-GIS systems implementing different, and often highly complex, proprietary solutions.

Although the straightforward method for representing geographic objects described earlier in our simple vector data structure has the benefit of being easy to understand, there are a number of problems with such a data format. Point entities do not present a major problem, and are often stored in a similar way to this in larger vector-GIS systems. This is due to the zero-dimensionality of the point, which means that points cannot cross or contain other entities, and so present few topological problems. On the other hand, the storage of lines and areas in the simple model is extremely restrictive.

Several limitations apply to the lines section of our simple example. The main restrictions are twofold: the connections between lines are not explicitly coded within the data; and there is no directional information recorded in the data. Recording whether lines connect, as opposed to cross or terminate without connection, is vital within network data. Direction within line data may be of considerable importance in the representation of rivers or drainage systems, where the GIS may need to explicitly incorporate flow direction into analyses. For example, if we were to attempt to assess the feasibility of early trade networks based upon river transport, we would need to consider the costs incurred in travelling up and downstream on any given river in the network.

The problems of recording adequate topological information for area data are rather more severe than those of line data. We have already illustrated how the relationships between areas are not always simple. Additional problems are encountered as a result of areas that share boundaries, and of the relationships between areas enclosed by others and their surroundings—so called 'island' polygons. In cases where geographic areas border one another, such as soil,

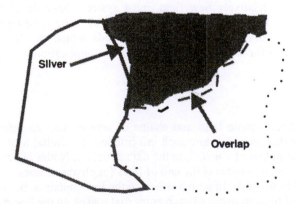

Figure 2.14 Areas represented by whole polygons showing some possible errors that result from shared boundaries.

geological or political maps, the simple model of area data (what is often termed the *whole polygon* model) repeats each shared boundary. This can lead to a variety of problems, particularly if data derives from digitised maps. It effectively doubles the storage requirement for each coverage, and therefore increases the time required to process the data, as two boundaries must be updated whenever a modification is made. In addition to this, at the data acquisition stage, the shared boundary lines may not have been digitised precisely. We will look in detail at the process of digitising in Chapter 3. These differences between the lines will show up as *slivers* or *overlaps* in the data coverage (see Figure 2.14). Additionally, there is no method of checking the data for topological errors such as figure-of-8 shaped polygons, where an error in the order of co-ordinates forces the area boundary to cross. These are frequently referred to in the literature as *weird* polygons and, as the name suggests, are to be avoided.

The simple whole polygon database also contains no record of the adjacency (or otherwise) of different polygons, so there is no way of searching for areas that border one another. Similarly, there is no straightforward way of identifying island polygons.

An enriched vector data structure

If analysis is to be carried out on data stored as vectors, the data must be capable of answering basic questions such as *are two polygons adjacent?* or *does this area contain any islands of different value?* Our brief examination of the simple data model should have illustrated that points, lines and whole polygons are not sufficient to represent both the spatial and topological complexity of real-world geographical data. As a result of this, more sophisticated data models are necessary in situations where either (1) analysis is mainly to be carried out on vector data themes or (2) the accuracy and redundancy problems associated with whole polygon data are deemed unacceptable.

The key to solving the problem lies in the incorporation of topological information explicitly into the data structure. As intimated earlier, there are unfortunately few data standards for the storage of full-topology geographic data. Because GIS is a relatively new field, most systems have developed largely in isolation and have adopted slightly different ways of storing and representing data. The result is a plethora of proprietary data standards, which, incidentally, can cause extensive problems for the porting of data from one application to another.

Networks

As implied earlier, simple lines and chains of arcs contain no information about connectivity or direction. To store such information in a digital form it is necessary to introduce the notion of a *node* into the data structure. Nodes are essentially point entities that are used to indicate the end of lines (or chains of lines), and to store the characteristics of the ends. In such a structure the line contained between two nodes is often referred to as an *arc*, and each point that makes up the line may be referred to as a *vertex*. Arcs that share a node are said to be linked.

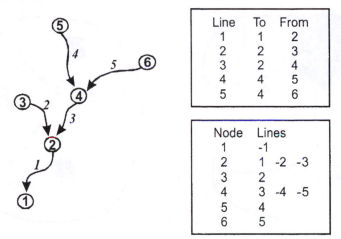

Line	To	From
1	1	2
2	2	3
3	2	4
4	4	5
5	4	6

Node	Lines		
1	-1		
2	1	-2	-3
3	2		
4	3	-4	-5
5	4		
6	5		

Figure 2.15 A simple river network represented by a list of lines (chain of arcs) and a separate list of nodes.

In the example in Figure 2.15, two tables have been used to represent a simple river network. The 'Node' table holds a record of which lines are incident at each node, and also which end of each line is incident at each node. This has been done by using the number of the line (*i.e.* '1') to indicate that the line starts at this node, or a negative number (*i.e.* '-1') to indicate that it ends at the node. A quick check through the nodes table should then reveal that each line number has one (and only one) negative equivalent so we can use this as a quick check: add up the numbers in the lines section, and it should come to zero. If not, then we certainly have an error. This data structure now represents the direction of each line, which in turn could represent the direction of flow within the river system.

Note that the 'from' and 'to' fields in the lines table are not strictly necessary, because they can be deduced from the nodes table. However, many systems include them to speed up processing. Note also that no *geographic* or *attribute* information has been included in this example. As in the simple data structure presented earlier, these must be stored but they have been left out here for the purposes of clarity.

Area data: arc-node-area data structures

We have seen some of the limitations of simple polygons as a model for storing area data. These have been summarised by Burrough (1986:27) as:

1. Data redundancy because the boundaries of polygons must be stored twice
2. Resultant data entry errors such as slivers, gaps and weird polygons
3. Islands are impossible to store meaningfully
4. There is no way of checking the topology of the data for these errors.

Fortunately, the arc-node model of network data outlined above, can be extended to record area data. The following discussion describes a simplified example of an arc-node structure, loosely based on the Digital Line Graph (DLG) format of the US Geological Survey.

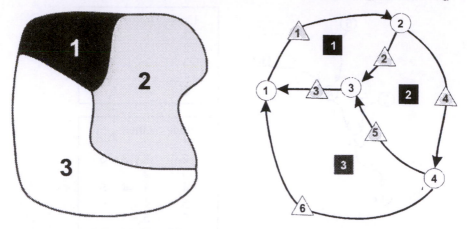

Figure 2.16 A simple topology problem. Showing (left) three areas of value 1, 2 and 3 and (right) a representation of their topology. Line numbers are given in triangles, nodes in circles and areas in the squares.

Figure 2.16 (left) shows a simple example of a configuration of areas that might require encoding. This shows three areas of value 1, 2 and 3, which have shared boundaries. To overcome the shortcomings listed above, what is required is a data model that records the boundaries only once, and preserves the relationships between the areas. As a first step we might give the lines and nodes numbers, and record these as for network data. If the lines and nodes are numbered as in Figure 2.16 (right), a 'first stab' at encoding Figure 2.16 might be as shown in Table 2.2.

Note that Table 2.2 deviates slightly from the network example in that the lines do not record their start and end node (as we said, this information can be deduced from the nodes table) and also that some spatial data has been included (the co-ordinates defining each vertex in the overall line or arc). All the attribute data has once again been left out for the sake of clarity. As before, the lines have been given a direction by recording each as a positive number at the start node and

Table 2.2 Introducing nodes into the data structure.

Line	Co-ordinates defining vertices
1	$x_1, y_1 \ldots x_n, y_n$
2	$x_1, y_1 \ldots x_n, y_n$
3	$x_1, y_1 \ldots x_n, y_n$
4	$x_1, y_1 \ldots x_n, y_n$
5	$x_1, y_1 \ldots x_n, y_n$
6	$x_1, y_1 \ldots x_n, y_n$

Node	No. lines	Lines
1	3	-6, -3, 1
2	3	2, 4, -1
3	3	3, -5, -2
4	3	6, -4, 5

as a negative one at the end node—again this is a common way of allowing the system to quickly check the topology.

In order to fully record the area component of the data, however, we need a third table. Although there is strictly no real need for the area data to be linked directly to any spatial entities, area codes are often recorded by creating special points within the relevant areas and attaching the area data to these. There will be three area entities for this example. We proceed by recording which area is to the left and which to the right of each line, and adding this to the lines section as shown in the top of Table 2.3. A third data section (Table 2.3, bottom) now records the data about each area and we now have somewhere we can put the attribute data for each area (although it is not shown here). This new section records which lines form each area's polygon, coding them as positive if they go clockwise around the polygon or negative for anti-clockwise. As before, this can actually be deduced from the left and right areas of the lines table and thus represents a compromise between data redundancy and processing efficiency.

Island and envelope polygons

So far our example has been restricted to polygons with shared boundaries. In Figure 2.17 we have a much more complicated topological structure that includes an island (4) and a 'complex island' (inside area 3). Initially we can proceed as before, and assign numbers to the lines, nodes and areas as shown in Figure 2.17 (right). This time we have included a special 'bounding polygon' coded 0 (zero) to represent the area that surrounds our geographic regions, and a special line 0 (zero) to describe it. This is sometimes, but not always, required by digitising programs.

Table 2.3 Introducing areas into the data structure.

Lines	Vertices	Left area	Right area
1	x, y ...	4	1
2	x, y ...	2	1
3	x, y ...	3	1
4	x, y ...	4	2
5	x, y ...	3	2
6	x, y ...	4	3

Nodes	X	Y	No. lines	Lines
1	?	?	3	-6, -3, 1
2	?	?	3	2, 4, -1
3	?	?	3	3, -5, -2
4	?	?	3	6, -4, 5

Areas	No. lines	Lines
1	3	2, 3, 1
2	3	4, 5,-2
3	3	-5, 6,-3

The line and node tables are constructed as for the previous example, including these extra lines and nodes. These are shown in the three tables of Table 2.4. The areas table (*i.e.* the information associated with the label points) is now modified so that each area record also includes the codes for any areas it encloses (*islands*), and for any area that encloses it (*envelopes*). In a similar way to the direction of lines, islands are recorded as positive numbers while envelopes are given as negative numbers. In this way the entire hierarchy of islands and envelope polygons can be recorded within the areas table.

These three tables now represent a full description of the area, including the geographic and topological elements. More importantly, this structure does not repeat any of the lines—so is not prone to the problems of 'whole polygon' structures—and it allows a vector-GIS to quickly find the logical relationships between all the areas of a spatial theme. Attribute data can be linked to the area identifiers as for our simple data structure summarised in Figure 2.10.

Although vector-GIS vary considerably in their precise data formats, the basic principles are similar to those described here, with a full geometric description of the geographic and topological components of the data being built up from the relationships between points, lines and areas.

Summary

Having traced the various table relationships around Figure 2.17, you should by now appreciate the complexities faced in calculating and storing topological information for area data. To use and understand the geometric relationships encoded on a map sheet involves a relatively trivial visual task, one that we perform so routinely we tend to take it for granted.

For the GIS to gain a similar understanding is far from trivial. To really understand what makes topological data structures different from the simple models discussed previously, try to think how you could create automated methods

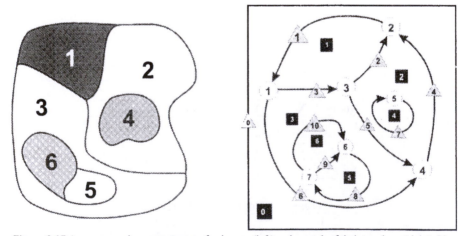

Figure 2.17 A more complex arrangement of polygons (left) and a graph of their topology (right). Line numbers are given in triangles, nodes in circles and areas in the squares.

to answer questions such as: '*is area 1 adjacent to area 2?*' or '*how many square km is area 3?*', without looking at the pictures! These types of questions can only be answered, by a computer program such as a GIS, if the vector data structure employed records the full complexity of the topological relationships present between the constituent geographical objects.

The point of these rather detailed examples is to illustrate the care required to store spatial information in a comprehensive and meaningful way. Fortunately, most of the work of creating this topological information and then checking that the logical structure of the database is coherent can be automated within the data input subsystem of the GIS. For example, in systems such as Arc Info or Cartalinx it is as straightforward as issuing the commands *build* or *clean* on a given vector data layer, or selecting the *create topology* button on a menu bar.

Table 2.4 The topology of Figure 2.17. Lines numbers are given in triangles, nodes in circles and areas in the squares.

Lines	Vertices	Left area	Right area
1	x, y ...	7	1
2	x, y ...	2	1
3	x, y ...	3	1
4	x, y ...	7	2
5	x, y ...	3	2
6	x, y ...	7	3
7	x, y ...	2	4
8	x, y ...	3	5
9	x, y ...	6	5
10	x, y ...	3	6

Nodes	x	y	No. lines	Lines
1	?	?	3	-6, -3, 1
2	?	?	3	2, 4, -1
3	?	?	3	3, -5, -2
4	?	?	3	6 , -4, 5
5	?	?	2	-7, 7
6	?	?	3	-9, 8, -10
7	?	?	3	-8, 9, 10

Areas	No. lines	Encloses (+ve) Is enclosed by (-ve)	Lines
7	4	1, 2, 3	0, -1, -6, -4
1	3	-7	2, 3, 1
2	4	-7, 4	4, 5, -2, 7
3	6	-7, 5, 6	-5, 6, -3, -10, -8
4	1	-2	7
5	2	-3	9, 8
6	2	-3	10, -9

The user then need only remember a few basic rules while digitising a map in order to ensure the complete data structure is maintained. Different GIS digitising programs do this in different ways, and some are better and easier to operate than others. We will discuss the digitising process in more detail in Chapter 3. As with data standards generally, there is at present no one accepted method of digitising and storing this type of data. The worked examples given here present a simplified form of the data models that are employed by many GIS and Spatial Data Builders such as Arc Info, Cartalinx, SPANS, GIMMS or MapInfo. Not all proprietary data structures incorporate all the levels of the model, while some include extra information in each of the tables, building in a degree of data redundancy to speed up subsequent data processing. We firmly believe that to construct coherent and robust vector data layers it is important to realise the complexity of the task the GIS has to perform when we blithely select the option 'create topology' from a pull-down menu.

2.8 RASTER DATA LAYERS

Rasters are relatively simple data structures and raster data is used in a far wider set of contexts than simply GIS. Examples of other kinds of raster applications include geophysical survey software, photo retouching programs and scientific image processing packages. Raster structures store graphical information by representing it as a series of small parts or elements. In the image processing literature, these are referred to as *pixels* or sometimes as *pels* (both derive from a shortened form of the words 'Picture Elements'). In the GIS literature these units are frequently referred to as 'grid cells' or simply 'cells'.

In a raster system, the spatial database once again comprises a series of georeferenced thematic layers. Unlike the vector data model, in raster systems we do not define individual features (whether points, lines or areas) within each thematic layer. Instead, the study area we are interested in is covered by a fine mesh of grid cells—Gaffney and Stancic (1991:26) have likened it to a chess-board—and each cell is coded on the basis of whether it falls upon a feature or not.

As a result, a simple raster representation of an 'X'-shaped conjunction of field boundaries might look like Figure 2.18 (left). Here the junction is represented in terms of a matrix of positive values for the cells that fall upon a boundary and zero values for those that do not. A simple text file to store that data (shown on the right) might therefore consist of a series 1s and 0s, with commas as dividers to separate the cells from one another.

Some file structures (for example the Idrisi ASCII storage format) are almost as simple as that shown in Figure 2.18 (right). Normally text files are not used to represent the pixels, instead the storage is based on individual bits and bytes. This is because raster data files can contain extremely large numbers of cells. Consider, for example, a region of 20km by 20km where each cell represents a 100m x 100m land parcel—even at this course resolution the data layer will contain 40,000 individual cell values.

Returning to the example of the vector data structure, raster systems represent the three fundamental mappable units in the following way: points by a series of single grid cells; lines by a set of connecting cells; and areas by contiguous blocks of adjoining cells.

1,	0,	0,	0,	1,
0,	1,	0,	1,	0,
0,	0,	1,	0,	0,
0,	1,	0,	1,	0,
1,	0,	0,	0,	1,

Figure 2.18 A simple raster (left) and corresponding data file (right).

Attributes and raster structures

In the vector model, the spatial representation of features and their non-spatial attributes are kept separate, as a series of separate text files or a linked relational database. In the raster model the representation of features and their non-spatial attributes are instead merged into a unified data file as attributes are encoded directly into the thematic layer. It should be noted that this situation is changing as a number of raster systems (for example Idrisi) begin to support some degree of attribute file or external database access. In such systems rather than an attribute value, a numeric identifier is encoded into each cell that acts as a link to an external attribute database file in a similar way to a vector identifier label. This type of raster layer has been referred to as a *data frame*.

Despite such enhancements, it is fair to say that in the raster data model we are recording not a series of discrete spatial features, but the behaviour of an attribute in space. The particular attribute value of the ground surface we are interested in is recorded as the value for each cell. The numeric value assigned to each grid-cell can correspond to: a simple presence-absence (1=archaeological site, 0=no-site); a feature identifier (*e.g.* 1=woodland, 2=urban, 3=lake); a qualitative attribute code (soil type 1, 2, 3 *etc.*); or a quantitative value (73m above sea level).

Locating individual cells

In terms of georeferencing a raster thematic layer, in practice the left, right, top and bottom co-ordinates of the grid-mesh are recorded, as is the resolution (*i.e.* dimensions) of each grid cell. The locations of individual cells are determined by referencing each cell according to its position within the rows and columns of the grid. For example, if the grid cells are 30m square (*i.e.* the grid resolution is 30m) and the corner co-ordinates of the overall mesh are known, it is a trivial task to count the cells along and up/down from any corner adding/subtracting increments of 30m to locate any one cell.

Topology and raster structures

It will no doubt come as some relief to note that unlike the vector model, there are no implicit topological relationships that need to be encoded within raster data. We are after all not recording individual spatial features but the behaviour of a particular attribute in space.

Sampling and resolution

The feature that most distinguishes raster approaches to GIS data from vector ones is the issue of sampling or *quantization,* which occurs because real space is continuous while its computerised representation is discrete. Moving between the two can cause many problems and can generate 'artefacts' that have nothing whatsoever to do with the properties of the real world we are trying to represent.

The choice of grid cell resolution is critical to any analysis we may want to undertake. The finer the resolution, the closer and more detailed the representation is to its ground state. There are no rules to guide us, only the requirements and tolerances of the analyses we would like to perform. It is obvious that the junction of field boundaries depicted in Figure 2.19 (left) is not a very good image of what are probably the narrow, continuous lines of the original. Of course, the representation could be improved by increasing the number of samples that are taken, in other words increasing the *resolution* of the data. The junction in Figure 2.19 (right) shows the effect of increasing the number of pixels along each axis of the image by a factor of 2. Instead of the mesh being 5 pixels deep and 5 wide, the image is now 10 pixels deep and 10 wide. This new 10 x 10 image is still not particularly good, particularly at the junction itself, but it is rather better than the 5 x 5 image.

Ideally, we would want to have as high a spatial resolution as possible. Unfortunately, increasing the resolution has a significant impact on the storage requirement of the resultant data files. If the cells of our example land parcel were 20m by 20m (instead of 100m x 100m), then we would need 1000 x 1000 =1,000,000 cells in the data file rather than the original 40,000. This will result in a file around 1.5 million bytes in size, or around 1.4 Megabytes.

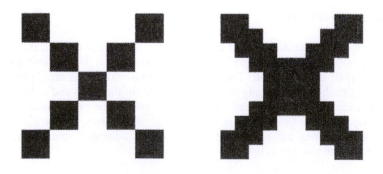

Figure 2.19 The effects of doubling the resolution of an image.

STORAGE REQUIREMENTS FOR RASTER DATA

One of the greatest problems with raster files is the large amount of storage space that they require, so predicting storage requirements is very important. To understand how to predict the storage requirement for a given application, it is first necessary to understand a little of how numbers and data are stored within computers.

At the lowest level, computers store information in *registers*. These are capable of representing only two values: 1 or 0. Using two of these registers (called bits), however, we can store four different binary numbers as follows: 00, 01, 10, and 11. It turns out that n bits can represent 2^n different numbers so that 4 bits can store 2^4, or 16 values and so on. One byte (eight bits) can store 2^8 or 256 different values or the numbers 0 through 255.

Because modern computers have very large memories and disks (at least compared with a decade ago) it is now common to refer to Kilobytes (Kb), Megabytes (Mb) or even Gigabytes (Gb) of memory. There are 1024 bytes in a Kilobyte, 1024 Kb in a Megabyte and 1024 Mb in one Gigabyte and so on (1024 is 2^{10}). Equipped with this arcane knowledge, we can begin to predict storage requirements for any given raster. We need to know two things:

1. the number of cells will there be in the raster;
2. the range of possible values we need to store in each cell.

The first is easy: we can calculate this from the size of our area, and the size of the individual cells. If we were interested in a region of 20km by 20km, and each grid cell represents an area of 100m by 100m, then the region will be represented as 200 cells (200 x 100m = 20,000m) wide by another 200 high. This results in 200 x 200 = 40,000 cells.

Next, we need to understand the range of possible values. This defines the storage allocation that needs to be made *for each cell*. If the data theme will only consist of two possible values—for example 'presence of an archaeological site' and 'absence of site'—then technically we need only allocate one bit for each cell. Usually, however, the cells can have many different values, as is the case in remotely sensed data, so that a larger unit of storage needs to be allocated to each cell.

Let us assume that our 20km by 20km data theme is a remote-sensed image, in which the cell values are scaled from 0 to 255. In this case, each cell needs to be stored not as one bit (2 possible values) but one byte (256 possible values). From this we can work out the storage requirement for this raster—the number of cells multiplied by the storage for each cell. In this case, the data file will be 40,000 bytes, which is a bit less than 40 Kb. Increasing the number of possible values that each cell can contain obviously increases the storage requirement for the cell. For example, if our image contains values that can be distinguished more finely, perhaps scaled between 0 and 4000, we will need to allocate more bits per cell—in this case 12 bits will allow values between 0 and 4096. Now the data file will be 40,000 x 12 bits = 480,000 bits or 470kb. Fortunately, most systems allow large raster files to be *compressed* (see main text) so that they do not, in practice, take up as much disk space as these calculations would suggest.

Compression

While increasing the resolution of a raster grid to say 10 x 10 or 100 x 100 improves the approximation to ground truth, it also dramatically increases the storage space required and increases the redundancy of much of the information in the image file. Unlike vector systems, raster structures explicitly store the absence of information as well as the presence. For example, if we return to the 4 long barrows outlined in Figure 2.10, in a vector structure these would be represented by a very compact set of 4 identifier codes and co-ordinate pairs.

In a raster system we would have to cover the barrow study area in a mesh and then encode not only a positive value for those cells which overlay a barrow, but zeros or negative values in all of the cells which do not—in effect non-data and the higher the resolution, the more redundancy there is likely to be. Each arithmetic increase in the dimensions of the image, results in a geometric increase in the number of pixels which must be stored: a 5 by 5 image contains 25 pixels, 10 by 10 contains 100, while a 100 by 100 image contains 10,000 pixels—so increasing the linear dimension by a factor of 10 actually increases the storage space required by a factor of 100. Obviously the resolution of the image therefore has a rather dramatic effect on the storage space the image requires.

These increases in the space occupied by raster images can be offset by using compression techniques. These encode the data in a more efficient way by removing some of the redundant data in the image. The simplest technique is referred to as run length encoding (RLE). RLE relies on the fact that data files frequently contain sequences of identical values which can be more efficiently stored by giving one example of the value, followed by the number of times it is repeated. For example, the 10 x 10 image in Figure 2.18 could be coded as pairs of numbers, in which the first represents the length of the run, and the second the value—so that 4,0 represents four consecutive zeros, reading right to left. This might look something like this:

```
2,1  6,0  2,1
3,1  4,0  3,1
1,0  3,1  2,0  3,1  1,0
2,0  6,1  2,0
3,0  4,1  3,0
3,0  4,1  3,0
2,0  6,1  2,0
1,0  3,1  2,0  3,1  1,0
3,1  4,0  3,1
2,1  6,0  2,1
```

The data file now only needs to store 68 numbers as opposed to the 100, which would be required for an uncompressed data file, providing a 'compression ratio' of about one third. Obviously little advantage is gained from compression where there are not enough identical values in long runs to make it worthwhile. In addition, a further disadvantage of compressed data files is that they must be decoded before they can be used, which takes additional time. For example, if the 5 x 5 raster on the left is encoded, the result is actually larger than the uncompressed file because the coding carries a data 'overhead' which must be recouped before any saving in storage can be made. In reality, raster grids are usually much larger than these examples, and sometimes contain a high proportion of repeated values. RLE encoding in images can therefore, depending on the context and on the picture, save considerable storage space.

Because the size of storage in a compressed data file depends on the *content* of the file as well as the dimensions, it is not possible to predict exactly the storage that will be needed for any given application. A good strategy is to obtain an approximation of the ratio, which might be obtained from compression by experimentation with typical data files, and then to use this to estimate the overall storage requirement.

Quadtrees

Some GIS (such as Tydac Technology's SPANS) utilise an alternative to raster storage that allows the user to specify a highly effective resolution whilst mitigating against the commensurate increase in storage space. This is termed a *quadtree data structure* (see *e.g.* Rosenfield 1980, Tobler and Chen 1993). It is based on a successive division of the stored region into quarters, so that the 'resolution' is determined by the number of successive subdivisions used. Values are stored in a hierarchical format, working downwards from the first quartering of the region (Figure 2.20). The advantage of quadtree data is that it can be more efficient than a simple raster format for the storage of some types of data, such as soil type or administrative boundaries. This is because some of the regions need only be subdivided a few times, while the full depth of the tree is used to provide high definition at boundaries. Burrough (1986) gives a fuller description of quadtree data structures, together with a more comprehensive illustration, but it is sufficient for our purposes to realise that quadtrees can be an efficient and elegant storage option. In saying this, it should be acknowledged that, like run length

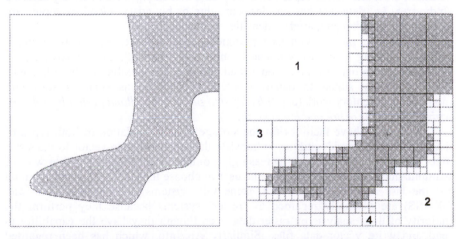

Figure 2.20 Quadtree data, to five levels (fifth not labelled). Some areas need only be subdivided once, while border areas require the full five divisions.

encoding, they require the system to do more work than for simple raster encoding, and so decrease the speed with which the GIS can process data.

Summary

Raster storage is simple but typically data-intensive unless the storage is at a low resolution, or mitigated through the use of compressed data or quadtree structures. Unless the resolution of a data theme is predetermined—for example where the cell samples derive from geophysical survey or fieldwalking—then a compromise is necessary between a high resolution, which provides better definition of features, and the increase in storage requirement and processing time that this generates. Analysis of raster data is relatively straightforward to undertake, although it is more limited than for vector data because the spatial relationships between geographical areas are not explicitly represented.

2.9 WHICH IS BEST—VECTOR OR RASTER?

A question that has often cropped up in archaeological discussions of GIS is *which is best — vector or raster?* As with most issues there is no clear-cut answer, with both approaches being adept at addressing different kinds of problems.

As we have discussed, unlike the vector model, which stores features as highly compact co-ordinate lists, the raster model is much more data intensive. However, the raster representation does define geographical space in a simple and predictable way — the spatial extent of a given attribute is always represented as a rectangular grid. As a result they are ideally suited to problems that require the routine overlay, comparison and combination of locations within thematic layers. As satellite and aerial-photographic data sources are themselves stored in a raster format, raster systems are good at handling and manipulating such data. In addition, the raster model copes better with data that changes continuously across a study area, *i.e.* data that has no clear-cut edge or boundary. Good examples are topography and changing artefact densities in the ploughsoil.

The vector model, with its topologically defined lines and areas, is much better at working with clearly bounded entities and network based analyses (*e.g.* shortest distance between points on a road system). The flexible database linkages make it ideally suited to database intensive applications such as sites and monuments inventory work (*e.g.* '*identify and plot all Woodland period burial sites with evidence for ochre and log coverings*').

Many GIS have facilities for the storage and manipulation of both types of data model, and provide utilities to translate data from raster to vector formats and vice versa. This is becoming increasingly common and offers a flexible way of manipulating spatial data, allowing the user to choose the most appropriate method for the task at hand. As a result, some GIS systems (for example Idrisi and GRASS) are generally regarded as 'raster' systems because they perform the majority of their analysis on raster data even though they have the capability to read and write vector data files. Similarly Arc/Info, which has been regarded primarily as a 'vector' GIS system, contains powerful modules for creating and manipulating data as rasters. It is also fair to say that a whiff of 'straw man' pervades this sub-section of the text insofar as the question itself is rapidly becoming historical. Most modern GIS have both raster and vector capabilities within them and all spatial databases will comprise a mix of vector and raster data layers.

2.10 A NOTE ON THEMATIC MAPPING

As mentioned earlier, the use of thematic layers makes a lot of conceptual sense, particularly in the management of data. Having discussed in some detail spatial data models, some additional benefits of such an approach should have become obvious. In a vector-GIS the thematic layer structure enables us to overcome any limitations that might arise as a result of having to adhere to specific topological types (point, line or area) in any given layer. In raster systems it allows us to overcome the inherent restriction of only being able to represent a single attribute in any given theme or image.

2.11 CONCLUSION

There are two spatial data models in widespread use within archaeology. Vector systems enable features to be represented using points, lines and areas, and provide a flexible link to related databases of attribute information. Raster systems offer a consistent grid-based mechanism for the recording of attributes over space. Any spatial database created within a GIS will contain a judicious blend of both vector and raster data layers.

The major implication this has for archaeology is that the representation of the spatial form and extent of any archaeological entity has to conform to one of these schema. With some objects, for example a single artefact find in an excavation trench, a single point or—depending upon the resolution—a single cell may be adequate. But what of the partial remains of an enclosure, defined by an interrupted ditch? Does it represent an area feature or block of contiguous cells? If so, how do we cope with the uncertainty and permeability of the defining boundary?

The scale at which data is acquired is also important. A 13th century AD Pueblo or Roman villa may be represented by little more than a dot on a map at a scale of 1:50,000, yet comprise a complex series of areas and lines when transcribed from an excavation plan, geophysical plot or aerial photograph at a scale of 1:500 or better. Problems such as the latter are not restricted to archaeological data — for example, whether a major river be represented as a line or area — and fall within the much broader topic of cartographic generalisation. Problems such as these will be discussed in more detail in the next chapter, when we begin to look in detail at issues relating to the overall design and management of an archaeological spatial database.

2.12 FURTHER INFORMATION

Most general textbooks on GIS cover the material we have discussed in this chapter, and we would recommend Burrough (1986), now substantially re-written as Burrough and McDonnell (1998) (see particularly Chapters 2 and 3). A very comprehensive discussion of spatial data and the spatial database can also be found in Worboys (1995) where many more methods of raster and vector storage are discussed. This also contains detailed discussion of database theory, fundamental spatial concepts, models for spatial information and data structures.

Guidelines for the design of archaeological spatial databases can be found in the AHDS *GIS Guide to Good Practice* (Gillings and Wise 1998), particularly section 4 '*Structuring, Organising and Maintaining Information*'.

CHAPTER THREE

Acquiring and integrating data

"... the process of checking and converting the maps to a format readable by GRASS proved to be extremely time-consuming, taking about one third of the total amount of time available for the research." (van Leusen 1993: 106)

In chapter 2, we described how a spatial database must provide for the storage and manipulation of the locational, topological, attribute and metadata components of our data. In this chapter we move on to describe the ways in which spatial and attribute data may be obtained and incorporated within a spatial database. We will describe the most popular types of spatial data, attribute data and topological data and the techniques we use to integrate them into the spatial database. As the opening quote suggests, the acquisition and integration of spatial data can occupy a significant proportion of any project. Van Leusen's estimate of one third of the project may seem high but is, in our experience, actually rather low. It is not unusual for the identification, purchase, digitising, quality control and documentation of data to occupy the vast majority of a project but without careful attention to this stage, any analysis will be seriously undermined.

3.1 SOURCES OF SPATIAL DATA

As discussed in the introduction, archaeologists have long realised the importance of spatial information and have worked hard to develop ever more effective and accurate techniques for its collection and measurement. In the course of our everyday practice we think nothing of:

- parcelling up landscapes into regular or irregular sample units for ploughsoil survey;
- carefully measuring and levelling in the locations of finds within an excavation trench;
- laying out baselines for recording earthwork features;
- setting out grids for test pits or geophysical survey.

Modern archaeologists find themselves using a bewildering array of techniques and equipment for recording spatial information, ranging from tapes, compasses and simple pacing, through to laser distance measures and high-precision Global Positioning Systems (GPS). Despite the variety of ways in which data can be collected it is reassuring to know that the spatial data resulting from our fieldwork will invariably take one or more of a very limited range of forms. Typically we could use the data to:

- draw up a scaled map or plan;
- produce a list of co-ordinate locations ready to be stored on record cards or within a database;

- in the case of techniques such as geophysical survey, produce an image of the measured response across the study area. Other techniques that result in such images are remote-sensing techniques such as aerial photography.

Our principal sources of spatial data can therefore be summarised quite simply as maps, co-ordinate lists and images (Table 3.1).

Table 3.1 Principal sources of spatial data in archaeology.

Spatial data source	Example
Maps	Traditional map sheets, excavation plans, hachure surveys
Co-ordinate lists	Finds catalogues, gazetteers, sites and monuments inventories
Survey	Total station, theodolite, tape or GPS
Images	Aerial photographs, satellite images, geophysical survey plots

3.2 SOURCES OF ATTRIBUTE DATA

A spatial database comprises both spatial and non-spatial (attribute) information. Attribute data is the information we have *about* the objects or phenomenon whose locations and spatial properties we have carefully recorded. Taking the example of a chunk of pottery found in the ploughsoil of a field, the spatial data component would consist of one piece of information, the X,Y co-ordinate of the findspot. In contrast there will be a range of attributes relating to the particular pottery find. A non-exhaustive list might include the fabric type, form, colour, weight, state of preservation, membership of a known type series, cultural affiliation, provenance and date. If we subsequently chose to take samples from the sherd for the purposes of quantitative characterisation study, the volume of attribute information relating to the sherd would no doubt increase dramatically. As this brief example illustrates, attribute data is frequently complex and is very dynamic, changing as more detailed analyses and studies are carried out.

The most common sources of attribute data within archaeology are monuments inventories and registers, fieldwork reports, record sheets, notebooks and gazetteers. A less obvious source is the keys and legends, which we find associated with maps, distribution plots and plans.

3.3 CLARIFYING THE RELATIONSHIP BETWEEN SPATIAL AND ATTRIBUTE

To illustrate the relationship between the spatial and attribute components of the archaeological record we can take a careful look at the data recovered through the

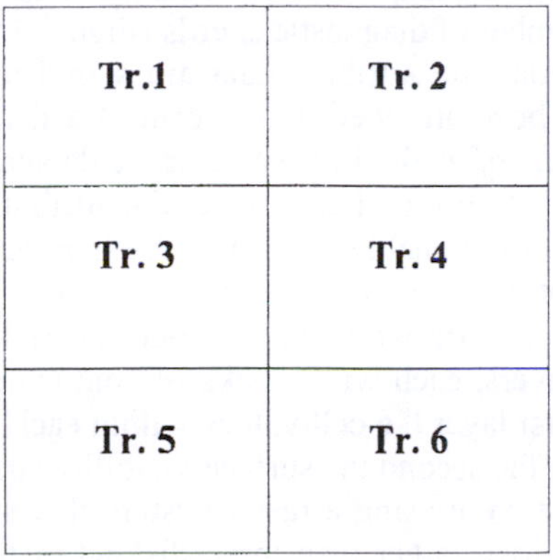

Tr.1	Tr. 2
Tr. 3	Tr. 4
Tr. 5	Tr. 6

Transect identifier	Ceramic count	Surface visibility (%)	Diagnostic pieces
Tr. 1	45	60	12
Tr. 2	42	60	8
Tr. 3	12	70	4
Tr. 4	9	70	3
Tr. 5	7	10	1
Tr. 6	3	20	2

Figure 3.1 An example spatial transect map (top) and collected survey attribute data (bottom).

process of fieldwalking. The following example is loosely based upon the data generated by the Boeotia project, which has undertaken an extensive programme of survey within an area of central Greece (Bintliff and Snodgrass 1985, Gillings and Sbonias 1999). In an attempt to study changing landscape configurations through time, the study area was parcelled up into a series of transect blocks. These could be irregular — defined by existing field boundaries, or comprise rectangular components of an overall sample grid. Within each transect, a percentage of the ground surface was visually examined for evidence of cultural material, primarily potsherds. These were duly counted. Along with the count, a sample of diagnostic material was collected and an assessment was made of the percentage of the ground surface that was visible to the walker. As a result we have a set of data comprising transects, ceramic counts, collected material and some assessment of surface visibility. The question we now face is how does this translate into the spatial and attribute components discussed above?

The spatial component of our fieldwalking exercise is straightforward, consisting of a series of discrete transects or areas. In the first instance the attribute information we have relating to each discrete transect comprises a total count, surface visibility estimate and number of collected sherds. Imagine we were to integrate this data as a vector layer. Our transects would be defined as areas or polygons, each with its own unique identifier label. Our attribute data would be structured into a simple database file comprising fields for transect identifier, count, visibility and number of diagnostic sherds (Figure 3.1).

Although the spatial and attribute data are stored in separate files, a direct link between them has been provided in the form of a shared piece of information (held in the *Transect identifier* field of the attribute database). The vector-GIS can utilise this shared piece of information to relate the attribute information directly to the spatial layer. This would enable us to plot the transects and colour code them according to the amount of pottery recorded within each.

In a raster data structure we would encode the fieldwalking data by using three discrete spatial layers, each with blocks of contiguous cells representing the transect areas. In the first layer the cell values within each block would correspond to the ceramic count, in the second the surface visibility and the third the number of diagnostic pieces. If we were using a raster system that supported the data frame concept, for example Idrisi, rather than three distinct raster data layers we would have a single raster layer through which we can visualise different attribute values. The cell values within each block in the data frame would correspond to the appropriate transect identifier code. These blocks of cells could then be related to a database of attributes in much the same way as in the vector example, and the data frame used to display or express the values of any selected attribute present.

We mentioned earlier the dynamic nature of attribute information. To illustrate this consider the situation where we may want to standardise the ceramic counts recorded on the basis of transect area. This would enable us to directly compare results between transects of varying size. We may also want to establish a correction factor for each standardised ceramic count, compensating for the surface visibility encountered in each transect. These new pieces of attribute information could be calculated and saved as new fields in the attribute database, perhaps entitled *area corrected count* and *visibility corrected count*. Equally, if we were to engage ceramic specialists to examine the diagnostic pieces collected from each transect we may be able to add further attribute information relating to the periods and form/functional types present in each transect. For an illustration of precisely this type of application see Gillings and Sbonias (1999).

3.4 INTEGRATING SPATIAL INFORMATION—MAP-BASED DATA

Having identified the main sources of spatial data utilised by archaeologists, we will now go on to look at the ways in which these classes of data can be integrated into the spatial database.

By far the most common means of integrating map data into a GIS is to digitise the map sheet. Digitising is the process whereby features are traced from a given map sheet into thematic vector layers stored within the GIS. The digitiser itself comprises a flat tablet linked to a puck that carries a crosshair and a series of buttons, which are used to transmit the location of the crosshair upon the surface of the tablet. A good way of thinking about the digitiser is to picture it as a Cartesian grid, with an origin in either the upper or lower left hand corner of the board. The units used to measure distance within the grid are specified by the sensitivity of the digitiser. Every time the appropriate button on the puck is pressed the digitiser sends the current location of the crosshair relative to the origin of the tablet. This is sent as an X,Y co-ordinate—in effect a point, the building block of all vector data primitives (points, lines and areas).

In terms of the actual practice of digitising, we affix the required map to the digitiser surface and then move the crosshair to each of the features we are interested in recording. With the press of a button we can then send the location of the feature to the GIS or digitising package. To enable us to input lines and areas, the majority of digitising packages allow the different buttons on the puck to be used to distinguish between: co-ordinate locations which signify simple point locations (vector points); nodes (the start and end points of lines); vertices (the straight line segments between pairs of nodes); and those indicating the positions of polygon labels (to allow us to allocate identifiers to discrete area features). In this way the full range of vector data primitives can be specified. It should be noted that some digitising hardware and software will also allow the tablet to *stream* information. Rather than only transmitting locational information upon the press of a button, when streaming a digitiser will send a continuous series of X,Y co-ordinates at a pre-determined time interval as the crosshairs are moved across the surface of the tablet. Although this effectively does away with the need to press any buttons, digitising in stream mode requires considerable expertise and a very steady hand. One area where it commonly finds application is in the digitising of long, sinuous linear features such as contours.

So far so good. We have our map affixed to the digitiser and can use the puck to inform the GIS whether we are recording discrete point, line, or area features. However, at present the co-ordinates we are recording are digitiser co-ordinates, measured in units such as inches or centimetres with respect to the origin of the tablet. To be able to utilise the data layers we are digitising in our spatial database we do not want to be working in arbitrary digitiser units but instead in the same

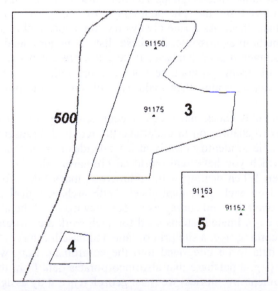

Figure 3.2 Hypothetical map sheet showing the locations of archaeological sites, the course of the major river in the study area and the boundaries of fields that have previously been fieldwalked.

real-world co-ordinate units used by the cartographers to produce the original map. A further step is clearly necessary to convert the co-ordinates from the digitiser's own internal system to the real world system we want to use. This process is referred to as a digitiser *transformation* or *registration* and serves to enable the GIS or digitising package to determine the precise series of multipliers and compensation factors it needs apply to automatically convert the incoming raw digitiser locations into real world co-ordinates. In practice, the process of transformation relies on the user being able to identify at least four points on the map sheet for which they can also provide real world co-ordinate information. The most useful points are grid intersections or the boundary corners of the map sheet, which usually have co-ordinate information either written on directly, or listed in the map legend. These points are variously referred to as *registration*, *control* or *tic* points in the literature.

To illustrate the process of digitising let us take a careful look at the various stages involved in digitising three layers of information from a simple map sheet. The layers comprise the locations of archaeological sites, the course of the major river in the study area and the boundaries of fields that have previously been fieldwalked. The sample map is shown in Figure 3.2.

Preparation

As with most endeavours, careful and thoughtful planning is the key to a successful digitising session. The first thing to consider is the map or plan itself. Is it printed on a stable medium such as Mylar film, or on paper? If paper, has it been folded so that it is hard to keep the sheet flat, or has it suffered any depredation such as stretching or tearing that may serve to compromise its spatial integrity? If you are using photocopies rather than originals you must also be aware of the possibility of distortions introduced by the photocopying process, particularly towards the margins of the map sheet. As a rule always try to use original map sheets printed on as stable a medium as possible. As these distortions may lead to less obvious errors in the resultant spatial database, if they are present it may well be worth abandoning the digitising programme until a more reliable source map can be found. If this is not an option then make careful note of the possible sources of error present.

It is also useful to check that adequate georeferencing information is provided on the map sheet to enable you to undertake the required digitiser transformation. You need to be able to identify clearly at least 4 (ideally 6 or more) points on the map sheet for which you have real world co-ordinates—the *control* or *tic* points mentioned earlier. When dealing with commercial maps this information will be carefully marked on each individual sheet. With archaeological plans and maps resulting from excavation and survey exercises they may well be absent. It is often the case that the co-ordinate systems used for such work are arbitrarily established at the time of recording (*e.g.* a site grid or tape and baseline survey). In such cases, if we want the data to be integrated into the co-ordinate framework of a wider spatial database (*e.g.* a database that also incorporates data from topographic map sheets and monument inventories) we must georeference it by assigning real-world co-ordinates to four or more prominent features. When selecting control points, it is sensible to identify an even spread of points around the perimeter of the area to be digitised. The corners of map sheets are ideal. Selecting control points bunched in one area of the map may lead to subsequent transformation errors in those areas furthest from the control point cluster. Having identified the points the next stage is to assign each of them a unique identifier code that can be marked upon the map, making a careful note of their real world co-ordinate values.

Once the control points have been identified, the next step is to decide upon the vector data layers you intend to create from any given map sheet. Once determined, you must then give each layer an unambiguous name. For example, if a layer is going to contain contour information derived from a topographic map sheet, a sensible name would be *fig3.2_contours*. As any GIS practitioner who has ever had to sift through 27 layers with variations of the name *DEMtry1* will tell you, clear, informative layer names are invaluable. Once you have decided upon the layers that are to be created and have assigned them names, the next stage is to decide whether each layer is going to contain point, line or area features. This may sound a rather straightforward task but often involves careful deliberation. For example, an archaeological site may be represented by a spread of surface lithic material. Do we digitise this as an area or simply indicate its position using the centre point of the spread? Equally a river can be represented by a single line (centre-line) or by each of its banks (area). The answer is typically dictated by the scale of the map we are digitising along with the type of analysis we ultimately want to carry out. Building upon this point, we have to appreciate that exactly the same kinds of question were faced by the cartographers who plotted the original map and it may be the case that a given map scale is simply not adequate for a particular type of study. For example, if we wanted to study the overall distribution of Roman marching camps in the North of England a 1:100,000 scale map, indicating the locations of the relevant features as discrete points, may well be adequate. If we then wanted to go on to compare the areas of a series of Roman marching camps in a restricted area of Northern England, we must accept that the 1:100,000 scale map is of very little use. Equally, if we attempted to digitise them from a map of scale 1:25,000, the areas depicted may already have suffered a considerable degree of simplification at the hands of the cartographers leading to errors in our area calculations.

Having decided upon the layer names, and the type of vector data (point, line or area) that is going to be input into each layer, the next stage is to establish a coding scheme for the feature identifiers that you intend to use. As discussed in Chapter 2, feature identifiers are crucial as they allow us to relate each individual vector feature in a given layer to the databases of attribute information we have about the feature. There are instances where unique identifiers are critical, for example to distinguish between archaeological sites or individual objects for which we have specific chunks of attribute data. Equally, there may be situations when shared identifiers can be used, for example all roads are given the identifier 1, all tracks 2, and all pathways 3. Before digitising it is essential that you construct a legend for each layer you intend to create. This should list the identifier codes used and the features they signify. An example of such a legend for the three data layers generated from the map shown in Figure 3.2 is given in Table 3.2. It is important to realise that when digitising area features all areas must have an identifier, even those that are created by default or refer to features not of immediate interest. For example, when digitising survey transects there may well be areas that correspond

Table 3.2 Sample identifier code legend for the map shown in Figure 3.2.

Numeric identifier	Feature type	Layer	Description	Notes
91150	Point	Fig_3.2_site	Fieldwalking scatter	Corresponds to project allocated site number
91152	Point	Fig_3.2_site	Fieldwalking scatter	Corresponds to project allocated site number
91153	Point	Fig_3.2_site	Fieldwalking scatter	Corresponds to project allocated site number
91175	Point	Fig_3.2_site	Fieldwalking scatter	Corresponds to project allocated site number
3	Polygon	Fig_3.2_survey	Fieldwalking survey block	Corresponds to project allocated block number
4	Polygon	Fig_3.2_survey	Fieldwalking survey block	Corresponds to project allocated block number
5	Polygon	Fig_3.2_survey	Fieldwalking survey block	Corresponds to project allocated block number
500	Line	Fig_3.2_hydro	River	Feature class
600	Line	Fig_3.2_hydro	Stream	Feature class
700	Line	Fig_3.2_hydro	Drain	Feature class

to modern towns, woodland or lakes that were not surveyed. All of these areas must be allocated identifiers.

Practicalities of digitising

Once you have undertaken the preparatory tasks described above you are ready to begin to digitise your map. As with GIS, there are a number of digitising packages available on the market. Some are embedded within GIS packages, for example the *Arc Digitising System* (ADS) module of *PC Arc Info*. Others comprise freestanding spatial data acquisition packages such as *Cartalinx*. Although they inevitably differ in the precise approach taken and overall ease-of-use and functionality, the basic steps outlined here are common to them all.

The first step is to fix the map to the digitising tablet, ensuring that it is as flat as possible so as to minimise any distortions. You then need to establish the digitiser transformation to be used by specifying the real-world co-ordinates of a number of points that can be precisely located on the map using the tablet crosshair. Once the points have been entered the digitising package will then determine how closely the digitised points fit the expected pattern of co-ordinate locations and report the magnitude of any discrepancies that exist. These can be caused by distortions to the map (as discussed) or poorly located control points. It will output this information in the form of an error measurement called the *Root*

Mean Square (RMS). This is expressed as a single value in the real-world units you are transforming to (*e.g.* metres or yards). To effectively use the quoted RMS it is important to acknowledge that mathematically it is the spatial equivalent to the standard deviation statistic. As a result, it is important to realise that a reported RMS value of 2.5m does *not* indicate that any subsequent point in the digitised data layer will be within 2.5m of its true location, but instead that the *average* error across the digitised layer is 2.5m. The *actual* error of any given point can vary across the image depending on the number, quality and regular placement of the control points used. At this point you must consider the RMS value and decide whether the error reported is acceptable for the type of data layer you are creating. Whilst you should always strive to get the RMS as low as possible for any given layer you should be prepared to accept a higher value for an old and well-used paper map than one printed on stable Mylar film.

It is worth mentioning at this point that most digitising systems give you the option of undertaking the transformation *after* the features have been digitised rather than before. The choice is up to the user, but if upon transformation you *do* find that you have an unacceptably high RMS error (due, for example, to stretching in the map sheet) it is better to be able to discard the map and find another more reliable source of data rather than have to discard the map and many days worth of digitising time.

As discussed earlier, whilst digitising you are in effect recording a series of X,Y co-ordinates that can be interpreted by the GIS or digitising package as:

- stand-alone point features;
- the start, end or intermediate points along the length of a line;
- area label points located within a series of lines which close to form an area.

Imagine you have a series of three lines representing trackways that share a common junction. In digitising the three lines you would want all three to share a single node location corresponding to the junction. Equally, when you are digitising areas, you want the start and end nodes to fall in exactly the same place so as to ensure closure of the area feature. Given the sensitivity of digitising tablets it is practically impossible for a user to place a node on exactly the same X,Y co-ordinate point as an earlier node. Fortunately, digitising packages enable us to overcome this difficulty by specifying what are termed *snap-tolerances* or *snap-distances*. A snap-distance corresponds to a zone located around each node, of a radius specified by the user. If another node is subsequently placed within this zone, the digitising package assumes that the user meant to place them in the same location and automatically snaps them together. There are no rules for how large a given snap distance should be, only a judicious blend of common sense and trial and error. Too large a radius and you may find that nodes that were meant to be distinct collapse into one another, too small and area features may not close and joined lines fail to connect. At the start of any digitising task it is worth experimenting to find a suitable tolerance.

Many digitising systems also allow the user to specify what is termed a *weed-tolerance*. When digitising a line, you begin by placing a node and then specify as many vertexes as are needed to adequately capture the shape of the line before ending with a second node. Straight lines require far fewer vertexes than curved ones. The weed tolerance is a user-specified distance that is used to remove unnecessary vertex detail from a given line. For example, if a weed tolerance of 5m is set the digitising software will reject any vertex co-ordinates that falls within 5m of the previous vertex point, thus simplifying the shape of the line. Unlike the snap distance, which should be readily modified and experimented with, in general it is advisable to accept the default weed tolerance offered by the digitising system as this represents a sensible compromise between excessive redundant detail and the adequate representation of features.

The importance of becoming familiar with tolerances cannot be over-emphasised as careful and sensible use will save an enormous amount of often very fiddly and time-consuming editing at a later stage.

As features are completed, for example the final node is placed at the end of a line, a point is located or label added to the centre of an area, the digitising package will prompt the user for an identifier code to attach to the feature. At this point the legend prepared earlier serves as a useful reference and aide-memoire.

The final stage in the process is to establish topology for the layer. This can be undertaken within the digitising software itself (*e.g. Cartalinx*) or within the host system (as is the case with *Arc Info*). It is at this stage that problems are commonly identified, for example area features not closing or labels being inadvertently omitted. All digitising packages or their host systems offer tools for interactively editing vector data layers to correct such errors. It should be noted that such editing tasks are often extremely complex and time consuming.

Scanning of maps and 'heads-up' digitising

An alternative pathway to integrate map-based spatial data sources into a spatial database is via the use of a flatbed or drum scanner. Scanners convert analogue input sources (in our case the map or plan) into a digital raster format. The file formats used by scanning devices are invariably designed for the storage of photographic images (for example TIFF and JPEG formats). Fortunately, these images can be readily imported into raster compatible GIS, spatially referenced and then converted into a raster-GIS data layer. The task of georeferencing the layer involves the identification of points on the image for which real-world co-ordinates can be provided. As with the selection of control points for the purposes of digitising, the quality of these points and their distribution within the image will have an enormous impact upon the accuracy of the referencing procedure.

The benefits of scanning over digitising are the fact that complex map sheets can be integrated in a fully automated fashion, without the time-consuming need to manually trace out features using a tablet and puck. If the aim is to simply provide a backdrop upon which archaeological information can be displayed then scanning offers an effective means of integrating map-based information.

If the scanned data is going to be utilised in the construction of thematic layers we have to be aware that the downside of the scanning process is that the data layers generated are completely undifferentiated. For example, take the map shown in Figure 3.2. It would be a relatively straightforward task to scan the map into an image file format and then convert this into a spatially referenced raster data layer that can be used as a base upon which other data layers can be superimposed. The layer created would record the presence (1) or absence (0) of an underlying feature for each cell in the raster grid. As a result we would know whether a feature exists at any point in the raster grid but would not be able to distinguish between cells which correspond to archaeological sites, the river or field boundaries. To be able to differentiate these features within the layer we need to manually edit the cell values for each of the feature types. With complex map sheets this can be a very time-consuming process, made all the more difficult by the wealth of information present on any given map, such as toponyms, explanatory text, and irrelevant detail, such as smudges and folds that have to be deleted from the final raster data layer.

A solution to this is to prepare either a cleaned version of the source map with irrelevant detail and extraneous information removed, or a series of thematic tracings that can be scanned in one at a time. In either case the process is time-consuming and once scanned, features will have to have identifier values attached to them. As a result an important decision has to be made as to whether the process will save any real time over conventional digitising.

A further consideration concerns the issue of resolution that we discussed in Chapter 2. Scanning devices vary in terms of their accuracy and resolution, which has a direct impact upon the degree of map detail that is captured on the resultant image. Use of a low-resolution scanner can lead to unwanted simplification in the final raster data layer.

One interesting technique of integrating map-based information into the spatial database relies upon a hybrid of the scanning and digitising processes often referred to as *heads-up digitising*. Here the map is scanned, read into the GIS and then displayed upon the screen. The mouse is then used to interactively trace features from the on-screen image into vector data layers—in effect digitising without the need of a tablet. To assist in this process, or in some cases fully automate it, a number of software tools are available which trace scanned lines to create vector data layers from scanned images. It should be noted that these tools are rarely incorporated as standard within GIS packages and are often expensive to purchase. In addition, the effectiveness of this process is once again a product of the prior processes of cleaning and tracing that have been undertaken on the map sheet prior to scanning (Gillings and Wise 1998:29). Whilst scanning offers a number of attractions over digitising, it can often be time consuming and very costly to apply.

3.5 INTEGRATING SPATIAL INFORMATION—CO-ORDINATES

The second of our major sources of spatial archaeological data comprises co-ordinate lists of the type commonly encountered in gazetteers, site inventories and regional monument records. A typical example is given in Table 3.3.

In terms of format, published co-ordinates should always follow standard surveying convention in being expressed in the form Easting: Northing. If you are working with raw archaeological survey data it is worth checking that this standard notation has been adhered to. Providing you can list the Eastings and Northings for a given set of features in a simple text file the GIS can read them to create the necessary spatial layer. When dealing with some national agencies you may find that co-ordinates are quoted not as discrete Easting and Northing values but instead as a single composite value commonly referred to as a grid-reference. To be able to integrate a location recorded in this way the grid reference must first be decomposed into its Easting and Northing components.

This latter point can be illustrated with reference to a location quoted using the Ordnance Survey national grid system for the UK. The following grid-reference locates a possible prehistoric megalith in the southwest of Britain: SU1164867982. To be able to integrate this into a GIS as a vector point the overall grid-reference must be separated into its 3 component parts. These comprise the two-letter prefix 'SU' and then the string of numbers split into two equal halves '11648' and '67982'. The two-letter prefix serves to identify the precise 100km grid square that the location can be found in, whilst the two numbers correspond to the Easting and Northing co-ordinates within that square. To convert the grid-reference into a national co-ordinate the 'SU' needs to be converted into the Easting and Northing values of the bottom left hand corner of the 100km square within the overall national grid and then added to the Easting and Northing values proceeding it. In this case the 'SU' references grid square 400,000:100,000 giving us a final co-ordinate value of 411648:167982 for the suspected megalith.

When discussing co-ordinate lists a crucial point to consider is the precision that is being quoted. For example, are the co-ordinates to the nearest metre or 10 metres? In the case of the regional Sites and Monuments Registers for the UK, it is not uncommon to find co-ordinate locations quoted to the nearest 100 metres. For example, our megalith would be recorded not as SU1164867982 but instead as SU116679—what is commonly referred to as a 6-figure grid-reference. If you are working with a spatial database that has been constructed on the basis of metre precision then to be able to integrate the co-ordinate information with the other layers of spatial information you will have to round-up the co-ordinates, in effect adding a level of spurious precision to the resultant data layer. The SU116679 becomes 4116: 1679 which can be rounded up to 411600: 167900. If this is the case you should carefully record the level of rounding that has been applied to each co-ordinate in a given layer.

Once grid references have been broken down, and co-ordinates have been rounded up to the required level of precision, the next stage is to input the co-

Table 3.3 The structure of a typical co-ordinate entry file and its implementation as a simple comma delimited ASCII text file.

ID	Easting	Northing
1	456770	234551
2	456683	234477
3	450332	235001
4	451827	234010

1,456770,234551
2,456683,234477
3,450332,235001
4,451827,234010

ordinate list into either a spreadsheet or simple text file. Commonly co-ordinates are used to refer to a single point location, and providing the co-ordinates are stored in the correct digital format it is a routine task in any GIS to read them in to create a layer of point features. Table 3.3 shows a typical co-ordinate input file based upon the format utilised by the *generate* command of Arc Info.

Note that a unique identifier value is associated directly with each discrete co-ordinate pair. Co-ordinate data is input into the GIS as a vector data layer and the identifier values provided are assigned to the vector features the GIS creates. Although the example shows point data, by following the formatting guidelines for the particular GIS you are using it is equally possible to input line and area data from simple lists of co-ordinates. Once the vector data layer has been created it can be incorporated directly into the spatial database or converted into a raster format as desired.

3.6 INTEGRATING SPATIAL INFORMATION—SURVEY DATA

Increasingly, archaeological researchers are undertaking their own primary surveys of both archaeological remains and their context. Although it is tempting to think of this as a kind of 'real-world digitising', this is misleading. In fact, the design and implementation of effective surveys in a way that both minimises and permits estimates of errors is beyond the scope of this book. For a recent introduction, the reader is referred to *e.g.* Bettess (1998) for a specifically archaeological text or to Schofield (1993) for a more general text on survey. In this section we will simply review two kinds of survey that are currently popular in archaeology and some of the implications for data storage.

Survey with total station

A total station is, in essence, a digital theodolite with an in-built electronic distance measurer (EDM), a calculator (or simple computer) and usually an internal or external data logger. The total station can be used to take very precise, and rapid, measurements of angle and distance and has become the workhorse of archaeological landscape and excavation survey since its introduction in the mid 1980s.

When undertaking a total station survey, the angles and distances measured by the instrument are used to generate 3D co-ordinate locations, either within the total station or through post-processing. In practice a network of intervisible survey points (stations) is established around the feature of interest and a primary station is identified. If known, this station is then assigned its real-world co-ordinate value in the relevant national mapping system. More commonly this information is not known, and instead an arbitrary co-ordinate value is assigned. Often this will be something like 1000,1000,100 to enable the survey to extend in all directions without generating negative co-ordinate values or 1000,5000,100 to avoid confusing x and y co-ordinates. The North or East axis of the co-ordinate system is then aligned on a second station (commonly referred to as a reference object or RO) and all locations are recorded with respect to the resultant co-ordinate system.

For an accessible introduction to the practical mechanics of total station survey interested readers are referred to Bettess (1998).

The result is a series of 3D co-ordinate points (Easting, Northing, Elevation) recording features and topographic information or, often, a combination of both. Some total stations also allow attribute data to be attached to the spatial data while it is being recorded. This allows the surveyor to code, for example, that points represent site boundaries or fence lines or even to attach descriptions or measurements to surveyed information. An example of topographic data collected using a total station can be found in the Peel Gap case study mentioned in Chapter 10.

At present the survey data generated can be said to be 'floating' or 'divorced' because, although the locations of features have been determined with respect to the arbitrary co-ordinate system, we usually have no idea where in the world the co-ordinate system itself is. If the data is to be integrated into a spatial database the first task that has to be undertaken is to 'fix' the co-ordinate system into the co-ordinate system that is being used for the overall database. This involves determining the real-world co-ordinate locations and heights of the primary station and RO station, and adjusting the co-ordinates collected with respect to them accordingly. Increasingly the task of 'fixing' survey stations into a wider co-ordinate system has fallen to another survey technology, GPS.

Global Positioning Systems

The Global Positioning System (GPS) is a global navigation system, which derives position from radio contact with orbiting satellites. Although still a relatively new technology, GPS receivers are sufficiently well established and portable to be useful for a wide range of archaeological activities, including regional survey and site-level survey, providing data whose accuracy can be of the order of tens of metres to sub-centimetre depending mainly on the sophistication of the equipment.

GPS positioning relies on twenty-four US Department of Defence satellites that broadcast radio signals to GPS receivers. Position is estimated by analysing the transmissions, and comparing the time that codes were broadcast with the time they were received to calculate distance to the satellites. A minimum of four satellite signals are needed for estimation of position although GPS receivers usually track several satellites at a time—usually either eight or twelve—for improved accuracy.

A GPS receiver can estimate a position to the order of 5-10m accuracy but there are several factors that prevent a single receiver from improving much on this. *Selective availability*, the deliberate degradation of the civilian signal by the US Department of Defence has now been turned off, but the system is still prone to *e.g.* timing errors, satellite position errors (caused by slight variations in satellite orbits) and radio interference. Some satellite configurations are also better than others for calculating positions, so accuracy depends on the number and position of tracked satellites. Single GPS receivers are useful for regional scale surveys, but of limited value for any more detailed surveying. Fortunately, there is a way to make GPS measurements of position more accurate, by using *differential* GPS.

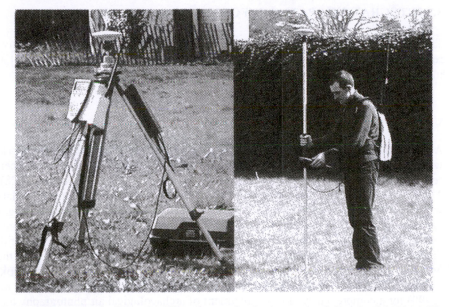

Figure 3.3 Survey-grade differential GPS equipment showing base station (left) and mobile (right). The circular antennae receive satellite information, while radio modems transmit differential corrections. (Equipment shown is Leica SR530, Department of Archaeology, University of Southampton, UK).

In a differential GPS, the surveyor uses two receivers. One is a portable receiver as before, whilst the other is used as a stationary base station. Data about the differences between reported and actual positions of each satellite can be recorded by the base station, and can be used to calculate the positional corrections that should be applied to one or more mobile receivers. The corrections can be saved to a file, so that the records in the mobile units can be corrected later—*post processed differential GPS*—or they can be broadcast to the portable unit every few seconds, allowing the unit to make corrections to its position as they are recorded, a technique called *real-time differential GPS*.

Some very specialised GPS systems perform extremely accurate timing of the radio signals by measuring the time taken for the radio *carrier frequency* to reach the receiver, rather than the special code it carries. This trick, combined with accurate differential corrections, allows survey grade GPS units to provide absolute accuracy to sub-centimetre levels suitable for surveyors (see Figure 3.3).

GPS systems can have sophisticated data loggers that permit data to be captured as points, lines or areas making incorporation within a GIS relatively straightforward and, as with Total station survey, many systems allow the surveyor to enter attribute data during the survey. Although not entirely unproblematic, GPS-based surveys do not suffer from some of the complicated inter-dependencies of conventional (theodolite or total station) surveys because each observation is essentially independent of the others. Obviously, one of the main considerations when integrating spatial data captured by GPS is to ensure that the accuracy and precision of the system is correctly estimated.

3.7 INTEGRATING SPATIAL INFORMATION—IMAGES

The final type of spatial data source we would like to discuss are those we have referred to as images. This rather broad term encompasses sources of spatial information as diverse as aerial photographs and geophysical survey plots.

Conventional aerial photographs

Aerial photography was pioneered in the years between the first and second world wars by O.G.S. Crawford, the UK Ordnance Survey's first archaeological officer, see for example the pioneering *Wessex from the Air* (Crawford and Keiller 1928). The technique has now become an accepted and widely used source for spatial data within archaeology. More recently, the potential of electronically sensed images has also begun to be recognized, either using sensors held beneath conventional aircraft or remotely sensed images obtained from earth-orbiting satellites. A full discussion of aerial photographic rectification and interpretation or of the processing of remote sensing data is outside of the scope of the present book. Readers who wish to pursue these areas in more depth are referred to Schollar *et al.* (1990) for a comprehensive formal treatment of archaeological air photography and 'prospecting' techniques, Lillesand and Kiefer (1994) for remote sensing and

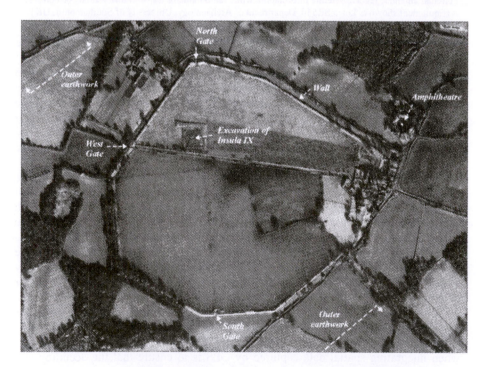

Figure 3.4 Ikonos image of part of the Roman town of Calleva Atrebates at Silchester, Hampshire, England. Image supplied by space Imaging (http://spaceimaging.com) ©2000. Reproduced with permission. Interpretation by Fowler (2000).

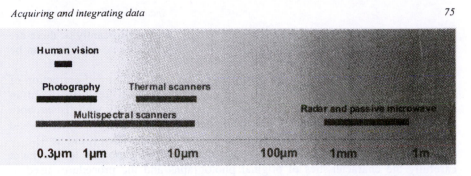

Figure 3.5 Approximate ranges of various sensors expressed in terms of wavelength (the scale is approximate, non-linear and interrupted where the outline is dashed).

image interpretation and to Bewley *et al.* (1999) for aerial photographic interpretation and rectification.

Aerial photographs frequently reveal archaeological features that cannot be seen, or at least appreciated fully, on the ground. It can reveal patterns formed by the shadows of slight earthworks (particularly in oblique light), or the different colours and textures of soil where, for example, chalky material is being ploughed into darker topsoil. Additionally, differences in the growth of vegetation over buried archaeological features can often cause 'crop marks' to be visible at particular times of the year.

Traditional equipment (*i.e.* chemical film cameras) can record either the intensity of the visible spectrum (monochrome film) or both the intensity and colour (colour film) within roughly the same range of wavelengths as the human eye (see Figure 3.5). It should be appreciated that this only represents a small part of the electromagnetic spectrum that is reflected or emitted from the land surface and which can be used to generate an image. Using conventional cameras, false-colour infrared film—so called because it is sensitive to wavelengths of light below the visible red range—can also be used to reveal features that many not be visible on normal film. Using filters placed over the lens, it is even possible to obtain several images of the same location recorded at different wavelengths.

Correction and rectification

However the image is obtained, it will not be suitable for immediate integration into the spatial database. All images will contain distortions caused by the different distances that the land surface is from the image sensor.

Aerial photographs are conventionally described as either *vertical* or *oblique.* Vertical photographs are taken under controlled conditions in which the height of the instrument and its orientation with respect to the ground are tightly constrained. As a result, the distortions that are introduced into the image can be predicted using the properties of the camera optics and a rectified image can be generated—an *orthophotograph*--that share some of the properties of a map.

Oblique aerial photography refers to images that are taken without ensuring that the camera is vertical with respect to the ground. In these cases, the distortions can be far more severe and the rectification of the image into an orthophotograph more difficult and error prone. Rectification of oblique aerial photographs is particularly problematical where the terrain height varies significantly or there are few reference points within the photograph for which reliable co-ordinates can be obtained ('ground control points').

As already mentioned, a full treatment of the rectification and correction of aerial photographs is beyond the scope of this book. A good treatment, from an archaeologist's perspective, can be found in Schollar *et al.* (1990) who discuss both the geometric transformation of archaeological air photographs (Chapter 5) and also many of the radiometric enhancements that can be undertaken (Chapter 4).

However, when establishing a spatial database it is important to document carefully the characteristics of original photographs and the procedures used to correct them. These should allow an estimation of the minimum and maximum errors, and the RMS error that is likely within a particular image.

Integrating the image

Providing that the photographic image has been rectified, the process of integrating an aerial photograph obtained by traditional chemical film methods into a spatial database is as follows. Firstly, a scanner is used to translate the paper print into a raster graphic file that can be read into the GIS. The resultant file comprises an image and little more, lacking any of the georeferencing information that is

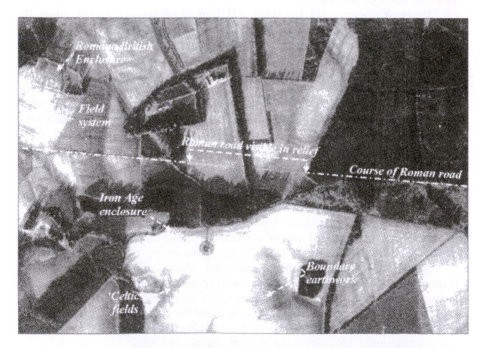

Figure 3.6 Part of a SPIN-2/KVR-1000 image of Farley Mount, Hampshire, England showing numerous archaeological features. Source: www.terraserver.com from SPIN-2 Imager. © 1999-2001 Aerial Images, Inc. and SOVINFORM-SPUTNIK. Reproduced with permission. Interpretation by Fowler (1999).

essential for its incorporation into a spatial database. The majority of GIS packages provide tools for spatially referencing such image files by allowing the user to match points on a given image to matching points on a previously georeferenced vector or raster data layer. To give an example, the GIS Arc Info has a command called *register* that initiates an on-screen interactive referencing system. Here windows are displayed showing the scanned photographic image and a corresponding vector data layer. Using the mouse the user can establish a series of links between features visible on the image and their counterparts in the vector data layer. Once four or more reference points have been identified a report of the RMS error incurred is given and the user can elect to use the information to establish georeferencing information for the image. As with digitising and scanning, the quality and distribution of control points directly affects the accuracy of the referencing procedure.

Once georeferenced the photograph can be treated as little more than a spatially referenced image for the purposes of display. The aerial photographic image can also be converted into a spatially referenced raster data layer for further processing and analysis.

Electronic scanners and remote sensing

Although it is possible to obtain chemical films sensitive to other parts of the electromagnetic spectrum (thermal bands, for example), these are unusual and it is far more common to use electronic instruments to measure the intensities of radiation outside of these ranges. These have several advantages over chemical film cameras in that they can be made sensitive to a far wider range of wavelengths, can be more easily calibrated, and they do not rely on an onboard supply of film that must be transported to the ground for processing. Additionally, electronic scanners can be designed to be sensitive to more than one band of radiation and to obtain readings for a range of wavelengths simultaneously, a process known as *multispectral scanning*. The main disadvantage of electronic scanners has been—at least until very recently—that they have been unable to approach the level of spatial resolution possible from chemical films.

Electronic scanners that simultaneously acquire measurements of hundreds of different bands are referred to as *hyperspectral* scanners and their use is described as *imaging spectrometry*. Because these instruments generate such a large number of discrete measurements across the electromagnetic spectrum, the reflections from each land unit sensed can be represented as an apparently continuous histogram and used to recognise and classify vegetation or mineral types by their spectral *signatures*.

Recently, the potential for satellite-borne instruments has begun to be realised within archaeology (Farley *et al.* 1990, Madrey and Crumley 1990, Cox 1992, Gaffney *et al.* 1996a) partly as a product of the improved spatial resolution that modern satellite sensors can generate but partly too because of the increased availability and reduced cost of these images. Satellite images can form a valuable source of data about the current environment. Using a combination of the bands that have been sensed, it is often possible to classify the image into urban, forest, pasture or agricultural land and this can be useful source data for comparison with the distribution and visibility of archaeological features.

Remote sensing from space

Remote sensing from space has only been possible for around 50 years. The earliest examples of images date to between 1946 and 1950, when small cameras were carried on V-2 rockets captured by the US during the Second World War. From c.1960 onwards earth-orbiting weather satellites (such as TIROS-1), began to be used to monitor cloud cover. Various photographic sequences were taken during the manned space flights of the 1960s and 1970s, and during the *Gemini* program, a series of nearly-vertical overlapping photographs of the south-western US (at a scale about 1:2,400,000) were taken and used to great effect to interpret tectonics, geomorphology and volcanology. The Skylab mission of 1973 produced over 35,000 images of earth with a 13-channel multispectral scanner and various other instruments (Lillesand and Kiefer 1994).

Today, satellite images are widely available. Commonly they are supplied either as prints that have been made by processing the satellite data, or as electronic data files supplied on tapes, disks or optical media. Images can normally be supplied as 'raw' data: virtually as sensed by the instrument, or with radiometric corrections made, or with both radiometric and geometric correction. Generally the corrected images will be of greater use within archaeology, as few organisations will have the necessary expertise or equipment to undertake the corrections themselves.

Images are now commercially available from an increasing range of instruments and satellites. In selecting which products may be useful for inclusion within an archaeological spatial database it is therefore important to understand the characteristics of the instruments, particularly in terms of their spatial resolution and the number and range of the spectral bands that they generate. We will briefly compare the instruments carried by three different groups of satellites—the six satellites of the US 'Landsat' programme, the French SPOT satellites and some more recently available data from declassified Russian and US satellite systems.

Landsat

Landsat is the name of a series of earth observation satellites launched by the United States. The first of these—Landsat 1—was launched in 1972 and there have been five subsequent successful launches (Landsat 2-5 and 7) and one unsuccessful launch (Landsat 6). The first three satellites have now been withdrawn (Landsat 1 in 1978, 2 in 1982 and 3 in 1983), as their sensors have effectively been rendered obsolete by subsequent launches.

It is important to understand that there is no such thing as a 'Landsat image'. Rather, the various satellites are different, and have carried a variety of different sensors each of which has produced images with different geometric and radiometric properties. In fact, four different instruments have been carried on the Landsat missions, of which the *Multispectral scanner* (MSS) carried by Landsat 1 to 5, the *Thematic Mapper* (TM) carried by Landsat 4 and 5 and the *Enhanced Thematic Mapper* (ETM+) carried by Landsat 7 are probably the most interesting for archaeological purposes.

The earliest of these, the Multispectral Scanner, produced data at a ground resolution of approximately 80m x 80m in four bands roughly equivalent to red,

Table 3.4 Spatial and radiometric properties of three of the sensors carried by Landsat satellites.

Sensor	Mission	Resolution	Band	Spectral range (microns)
MSS	1-5	c. 80m	1	0.5-0.6
		c. 80m	2	0.6-0.7
		c. 80m	3	0.7-0.8
		c. 80m	4	0.8-1.1
TM	4-5	30m	1	0.45-0.52
		30m	2	0.52-0.6
		30m	3	0.63-0.69
		30m	4	0.76-0.9
		30m	5	1.55-1.75
		60m	6	10.4-12.5
		30m	7	2.08-2.35
ETM+	7	same as TM	1-7	Same as TM
		15m	8	0.52-0.9

green, blue and infrared. The Thematic Mapper is a 7-band multispectral scanning radiometer whose resolution 30m in the 6 visible, near and short-wave infrared bands and 60m in the thermal infrared band. The 'Enhanced Thematic Mapper' (ETM+) carried by Landsat 7 is similar to the Thematic Mapper except that it is also capable of producing panchromatic images at a spatial resolution of 15m. Table 3.4 shows the parameters of each of these instruments for comparison.

SPOT

SPOT refers to the Système Pour l'Observation de la Terre, conceived by Centre National d'Etudes Spatiales (CNES) in France. The first satellite (SPOT-1) was launched in 1986 by an Ariane rocket but retired in 1990 to be used as a backup to subsequent SPOT-2 and 3, launched in 1990 and 1993. These are satellites orbiting at an altitude of 832km, and carrying identical sensor systems. The sensors consist of two 'High Resolution Viewers' (HRVs), each of which can operate either in panchromatic mode, sensing the range 0.51-0.73mm at a ground resolution of 10m or in three-band multispectral mode (0.59-0.59, 0.61-0.68, 0.79-0.89 mm) colour infrared mode at 20m.

The principal advantage of SPOT imagery over Landsat (prior to Landsat 7) has been the better spatial resolution of the images. Although only panchromatic, a ground resolution of 10m is sufficient to reveal large earthworks and even large-scale cropmarks.

Russian systems

Following the end of the cold war, the Russian government began to declassify data from various civil and military space programmes. This is now beginning to

Table 3.5 Examples of Russian satellite instruments, giving the size of the ground areas represented, spatial resolution and the number and type of the spectral bands (after Fowler 1995).

Sensor	Scene size	Resolution	Spectral bands
KVR-1000	40 x 300km	3-4m	0.51-0.76
KFA-3000	27 x 27km	2-3m	0.51-0.76
KFA-1000	80 x 80km	5m	0.57-0.81
MK-4	117 x 117 to 216 x216	6-8m	Multizonal: 0.635-0.69 0.81-0.9 0.515-0.565 0.46-0.505 0.58-0.8 0.4-0.7 Spectrazonal: 0.57-0.68 0.68-0.81
KATE-200	243 x 243 km	15-30m	0.5-0.60. 6-0.7 0.7-0.85
TK-350	200 x 300	8-12m	0.51-0.76

provide a thorough coverage of remotely sensed data and a real alternative to Landsat and SPOT. Many of the Russian satellite sensors are conventional photographic instruments rather than electronic sensors, and data is available either as prints or ready scanned. The principal advantage of many of these products is the claimed spatial resolution of the instruments: as good as 2 to 3m (claimed) for the panchromatic KFA-3000, for example or 6-8m for multi-band MK-4 data. Archaeological applications of this quality of data are already beginning to be explored (Fowler 1995, Fowler and Curtis 1995, Fowler 1999, 2001).

Others

A variety of other images previously classified as intelligence images are also now becoming available for general use. For example, three United States government satellite systems (codenamed CORONA, ARGON and LANYARD) were declassified in February 1995 providing more than 860,000 images of the Earth's surface, collected between 1960 and 1972. CORONA imagery has already been demonstrated by Fowler (1997) to have the potential to reveal archaeological detail. Any published account such as this of the available sources of multispectral data will inevitably suffer from rapidly becoming out-of-date, so the reader would be well advised to contact suppliers of imagery or online sources such as Martin Fowler's 'Remote sensing and archaeology' WWW pages (Fowler 2001).

Geophysical survey data

Before we move away from image data sources, it is worth mentioning a further source of data that is frequently encountered within archaeology. This is data produced by Geophysical Prospection (sometimes simply referred to as 'geophysics'). This term covers a wide range of approaches that involve measuring physical properties at the surface of the earth and using those to infer the structures found below.

Approaches can be based on the conduction of electricity (resistivity), magnetic properties (*e.g.* fluxgate gradiometry), or thermal properties. Some of the most promising approaches, but also the most difficult to interpret, are based on the transmission and reflection of electromagnetic pulse signals between two antennae called Ground Penetrating Radar (GPR).

In the majority of cases the result is, in effect, a flat map that can be integrated and georeferenced in exactly the same fashion as a vertical aerial photograph. Where these are available only as paper printout, they can be scanned, georeferenced and then treated as a spatially referenced image or converted into a raster data layer for further analysis and processing. More satisfactory is the case where the original data (or a pre-processed transformation of the data) are available for incorporation directly into the GIS using raster import functions. Most geophysical processing software will allow either raw or processed data to be saved either as an ASCII format, or as one of the popular image processing formats such as 'tagged image file format' (tiff) files, both of which are normally straightforward to import and georeference.

We do not propose to discuss in detail the processing or interpretation of geophysical images, as this is a significant and complex area in its own right. We would, however, caution strongly against the inexperienced using geophysical results as if they were unproblematic representations of archaeological features. The output from geophysical surveys requires considerable interpretation based on a sound knowledge of the physics and earth science involved. We would strongly urge, for example, that the practice of digitising features from rectified images ('heads-up digitising') should not be uncritically extended to geophysical data. Notwithstanding these issues, however, GIS does form an exceptional software platform for the integration and co-interpretation of geophysical survey data, where it can be directly compared with air photography, mapping data and any other source of information about the surface of the world.

The reader who wishes to know more should certainly consult Schollar *et al.* (1990) for a comprehensive formal treatment of resistivity, magnetic, electromagnetic (including GPR) and thermal prospecting methods. Alternatively, Wynn (1998) provides an overview of the main methods online.

3.8 INTEGRATING SPATIAL INFORMATION—EXISTING DIGITAL RESOURCES

Although some forms of spatial data, such as satellite images, have traditionally been available in a digital format, most other data sources have required some process of translation. We have discussed at length the most common of these, digitising and scanning. Increasingly spatial data sources such as maps are being made directly available in a digital, GIS compatible form. For example, many national agencies involved in the production of spatial data such as the US Geological Survey and UK Ordnance Survey have begun to offer digital map products. The important point we would like to emphasise here is that these digital resources have themselves been derived from one of the processes discussed above, *i.e.* digitising or scanning. As a result they are subject to the exact same list of potential sources of error and decision-making processes regarding feature representation, feature coding and layer naming as if the process were carried out by the user. As a result, when purchasing or acquiring digital data you should ensure that full details are supplied regarding coding schemes and factors such as RMS error. Integrating such sources into the spatial database is usually a very direct process as data is provided in GIS compatible formats or generic transfer formats that can easily be imported into the majority of GIS packages.

3.9 INTEGRATING ATTRIBUTE DATA

Up until now we have discussed in detail the major sources of spatial information encountered within archaeology and the ways in which these sources can be integrated into the spatial database. In the following section we are going to look at attribute information. As with the types of spatial data collected by archaeologists, sources of archaeological attribute data are numerous and varied. A non-exhaustive list might run as follows:

- paper-based card index;
- site and survey archives;
- published reports and journal articles;
- microfiche archives;
- typological databases or artefact type series.

Despite the variety of formats employed, all sources of attribute data take the form of some kind of *database*. To be able to integrate attribute data into the GIS this database must take a digital form. In the case of existing digital databases, this can be a very rapid process as the majority of GIS packages either read directly, or import, the most common database formats (for example dbase III and IV). In such cases the principle concern for the user is to ensure that there is a field for each record in the database that contains and matches the data type of the unique identifier values used to label the spatial features within the spatial data layer.

With non-digital data sources the process is more complex, as the existing records are translated and input into a digital database system. A full discussion of the mechanics of digital database design and implementation is well beyond the scope of the current volume and interested readers are referred to Ryan and Smith (1995), Date (1990) or Elmasri and Navathe (1989). With paper-based card indexes and microfiche archives this process may involve little more than the direct replication of the existing database structure in a digital medium. With reports, gazetteers and published articles the data may be far less structured and the process may well encompass two distinct stages. In the first the required pieces of thematic information are identified and extracted and in the second a digital database is established to store and manage this information.

In the case of single pieces of attribute data it may not make sense to construct a formal database. A good example would be the elevation values associated with a series of spot-heights or topographical survey points. In such cases rather than construct a full database, a simple ASCII text file can be established instead. This would consist of the identifier code of each sample point followed by a comma or space and then the relevant elevation value. These simple

ASCII attribute files can be read routinely by both raster and vector GIS packages. In saying this, for the purposes of effective and flexible data management and organisation, attribute sources containing more than a single piece of information for a given spatial feature are best stored as a digital database.

3.10 DATA QUALITY

Up to now, archaeologists have shown remarkably little interest in the quality of the data within archaeological spatial databases. This is not particularly a GIS problem—in fact archaeology has probably paid insufficient attention to the quality of information in *any* databases—but is brought into sharp relief by the use of GIS for several reasons.

Errors in the spatial component of the data become readily apparent when plotted or visualised. This is not the case with 'traditional' archaeological databases such as SMRs because the contents are rarely, if ever, converted into a visual representation.

GIS-produced visual products and reports can carry considerable 'authority' in a published article or report, despite the fact that the quality of the data and the inferential processes that underlie these products are frequently not discussed at all. The ability of GIS to generate new data themes means that poor data quality in source data can be propagated to the products in unpredictable ways. The most obvious example of this is the interpolation of elevation models from contours: it is far from clear exactly how the accuracy or precision of the contours affects the quality of the generated elevation model.

For these reasons, we need to pay some attention to the nature of data quality, and to the ways in which it can be controlled and expressed.

Accuracy and precision

Two of the most widely confused concepts relating to data quality are *precision* and *accuracy*. Although they are related, it is usually accuracy that is meant when either of these is used. In fact, accuracy refers to how close a value is to the true value. Although the accuracy of data can be very difficult to quantify—because we do not normally know the true value except by our data—it is usually estimated and given as a *standard error*. Precision, on the other hand, refers to the way in which the data is measured or stored. The finer the unit of measurement that can be resolved by the instrument, or stored by the computer, the more precise the data is said to be. Often, precision is used to refer to the number of significant digits that are used to measure, record or store data: grid references, for example, may be measured to the nearest one metre, or to the nearest ten metres, while survey instruments might record co-ordinates as centimetres, millimetres or even fractions of millimetres.

It can be seen, then, that accuracy and precision are not the same: it is quite possible to measure very carefully the location of something, but to measure it incorrectly and therefore produce highly precise but highly inaccurate information. At the same time, it might be wholly appropriate to cite a grid reference that is correct to the nearest 10 metres, and in that case the information can be entirely accurate but have a low precision. Of course, the two are related at least to the extent that the precision of the data must be sufficient to allow accurate recording: it is no use surveying a trench to 10m precision, or storing fractional data in fields that are only suitable for whole numbers.

It is also important to distinguish between relative and absolute accuracy when recording or integrating spatial data. A good example of data with a specified *absolute* accuracy is data derived from GPS recording. This is measured with reference to the positions of the GPS satellites, and although variations in satellite geometry or other conditions might mean that the accuracy of each measurement is different, each measurement taken is independent of each of the others. This means that the error quoted for each GPS observation is an absolute error and this is important because it follows that the error in the measured distance between two locations will always be the same whether the two points are within a few metres of each other or far apart—even if they are in different parts of the world. By contrast, in conventional (*e.g.* theodolite or total station) survey, the accuracy of any recorded point is relative to the accuracy of the point from which it was recorded. This means that we may know to the nearest millimetre where every artefact within our trench is, relative to the base peg for that trench, but if we do not survey in the base peg of the trench with similar accuracy and precision, then this will not translate into the same accuracy relative to any other spatial data sources we may have.

Inaccuracies and lack of precision can take a number of forms. Often the values in our database are just as likely to be above their true values as below them, but sometimes this is not the case. In some situations, the database values differ from their true values in a more structured way. This is the case, for example, if a survey instrument is calibrated to read all the values too high or when particular fieldwalkers are better at identifying flint, while others are systematically more likely to find pottery. In these situations, the inaccuracy of the data is said to exhibit a *bias* and, if the bias can be identified, then it can be possible to make allowance for it or even to make corrections to the data.

Good data quality, then, is partly a matter of ensuring that the data we use is appropriately precise and appropriately accurate for the job we have in hand, which itself is predicated on knowing what the accuracy and precision of our data are.

Factors effecting data quality

Inaccuracies exist in virtually all databases and arise through a variety of processes. Some of these, such as mistakes during data entry, are obvious but others are not and archaeologists need to be aware of the factors involved if we are to make sensible inferences from our spatial databases. A few of the factors that affect archaeological spatial databases are as follows.

Currency and source

Much of the data that archaeologists deal with is not collected for the immediate purpose, instead being derived from observations or measurements intended for another purpose. Although archaeological 'reality' may not change as rapidly as, say, vegetation or population data it is still important to consider both the source of the data, and the extent to which it is still apparent. Good examples of poor data quality due to the currency of the data can be found in most Sites and Monuments Records, where the information about archaeological sites may derive from paper records that are years or decades old. In the time between the recording (possibly at the last visit to a site) and the use of the data, attributes of sites such as their state of preservation may have changed considerably or their interpretation may be entirely different in the light of newer evidence. Spatial data is also prone to become out-of-date because different surveys or projections may be used for contemporary mapping products than those that were used for the original records.

Good spatial databases will maintain a clear record of the source of all data themes, including information about who recorded it, when and why.

Measurement and observation errors

These occur in virtually all spatial archaeological data including plans, sections, fieldwalking, geophysical survey, GPS positions and traditional survey data. Errors and bias are introduced into nominal and categorical data when, for example, a fieldwalker moves a pottery sherd to the wrong location or mis-labels a bag. Later, an artefact may be identified as belonging to the wrong class of pottery or the weights may be incorrectly recorded. Measurement errors may arise when instruments are poorly calibrated or configured. Examples include poor calibration of geophysical instruments, resulting in incorrect or inaccurate measurements. Magnetometry instruments such as fluxgate gradiometers, for example, are notoriously difficult to correctly calibrate. Even when operator error is wholly eliminated, no instrument will return exactly the same value for each observation and so there is an element of residual measurement error that should be estimated and documented. Well-designed fieldwork recording methods can minimise and control for operator and measurement bias but can never entirely eliminate it. It is good practice to make notes about the accuracy of any measurements undertaken. One way of obtaining an estimate of the accuracy of a measurement is to repeat the same measurement independently a number of times, and then plotting the distribution of the resulting observations.

A well-designed spatial database will include notes about the estimated absolute accuracy and precision of each data theme included.

Transcription errors

These can occur when data is transferred from one format, often paper, to another, usually a computer. Data entry can be so tedious that even a very accurate and scrupulous operator will make mistakes as data is typed into the keyboard or coded with a digitising tablet. Where plans and maps are being digitised, then there is the additional problem that the original may become stretched or distorted so that the digital product differs from the paper product (which itself will contain measurement errors).

Transcription errors also occur when one data format is translated into another—for example the destination format may have a lower precision for co-ordinates than the source. Transfer of data from vector to raster or raster to vector formats is

Table 3.6 A list of the metadata you would need to record when digitising a layer of thematic
information from a map sheet.

Metadata	Why is it needed?
The projection system used to generate the map	So that you can ensure the spatial integrity of the overall spatial database by ensuring that all of the layers are derived from the same projection. Where projections do differ you can undertake the required re-projection of the data layers
The scale of the source map	Given the ability of the GIS to work at any scale the user selects, to ensure that data collected at a specific scale is not used at any scale larger. This is a procedure that would at best produce distorted results and at worst meaningless ones.
The medium and integrity of the map sheet	To help account for any RMS errors encountered
The map publisher and copyright details	There may well be restrictions imposed upon the digital copying and subsequent use and distribution of any map sheets. As a result it is important to record this information.
The RMS error encountered on digitising the layer	To provide a record of the errors associated with the data layer – there is little point undertaking an analysis to centimetre precision if a given layer has associated errors of 2.5 metres! As errors can become compounded during the course of the various analytical possibilities offered by the GIS it is crucial that all error sources and assessments are carefully monitored.
The control points utilised and their real-world co-ordinates	To provide a record of the georeferencing information employed.
The coding and legend schemes used to name layers and label features within layers	To provide a simple reference to the various codes utilised, codes that may well seem obvious when designated but in a years time may not be so readily apparent!

particularly prone to introduce errors and inaccuracies. When moving from vector
to raster, then the sample rate (the number of pixels) must be carefully chosen to
represent the complexity of the vector data otherwise detail will be lost. If
translating from raster to vector, then the algorithm used is of central importance in
determining whether the edges created will be meaningful in the context that we
wish to use them.

A clear 'audit trail' should be present for each data theme, showing how it was
digitised and giving full details of the source.

3.11 METADATA AND INTEROPERABILITY

In Chapter 2 we discussed metadata and offered the following rather cryptic
definition: metadata is *data-about-data*. Metadata is particularly important for any
situation in which we may want our data to serve more than one purpose because it
allows us to avoid the tortuous problems of creating and enforcing data standards.
Instead, if we document our spatial database correctly we can concentrate on
making our system *interoperable* with other systems. The distinction is essentially
that instead of trying to ensure that we collect the same data, in the same format as

all other archaeologists—which presents both epistemological and practical problems—we create the contents and format of our database in a manner that is appropriate for our own situation and then use metadata to describe this content and format carefully so that other GIS will be able to utilise it and understand its limitations.

In practical terms, a better definition of metadata would be the information that you and others need to know to be able to interpret and make sensible and appropriate use of the information in the spatial database. For example, building on the discussions in Chapters 2 and 3, to be able to effectively utilise a layer digitised from a map sheet we would at the very least need to know the information encoded in Table 3.6.

Looking at how we store this crucial information, two basic types of metadata information can be identified. The first comprises information that can most effectively be stored as an attribute of a given spatial feature. For example, in our discussion of co-ordinate data in section 3.5, the issue of how to integrate co-ordinates of varying spatial precision was raised. The solution was to round up the co-ordinates to the desired precision (*e.g.* to the nearest metre) which overcame the problem at the expense of introducing spurious levels of precision into the spatial database. One way to acknowledge, record and compensate for this is to encode the original spatial precision of each feature as an attribute value. The second type of metadata needs to be stored separately from the data layers, in a dedicated or spreadsheet file. Rather than external to the overall spatial database these files should be regarded as an essential component. Some GIS packages offer ready-made templates for storing metadata and very detailed guidelines for storage and recording exist (Gillings and Wise 1998).

To summarise, metadata is important because it enables individuals to interpret and make appropriate use of spatial databases and individual data layers created by themselves and others.

3.12 CONCLUSION

In this chapter we have identified the principal sources of spatial and attribute data that archaeologists utilise and have examined the most common ways in which these data sets can be integrated into a GIS-based spatial database. We have tried to provide some information about the majority of sources that archaeologists draw upon, but have concentrated on those (like map data and remote-sensed imagery) which are most closely associated with spatial technologies and whose acquisition and manipulation is likely to be less familiar to most working archaeologists.

Throughout the discussion, reference has been made to the important pieces of additional information, metadata, that need to be noted and recorded to enable individuals to make best use of their spatial database. Having designed, documented and populated our spatial database, we will now go on to look at how we can manipulate and exploit the data held within it.

CHAPTER FOUR

Manipulating spatial data

"... Archaeologists must deal with vast amounts of spatial information in regional work and GIS are designed specifically for the handling and manipulation of spatial data. Consequently, GIS and archaeology may represent an ideal marriage." (Kvamme, preface to Gaffney and Stancic 1991:12)

As we discussed in Chapter 1, the feature that most clearly distinguishes GIS from other computer systems such as cartographic or drawing software, is the capability to transform spatial data and to use a range of existing spatial data to produce new data themes. This functionality allows users of GIS to model the processes that take place in the real world. It is often said of the spatial database in a GIS that the *'whole is more than the sum of the parts'*. What this rather cryptic statement means is that when using GIS, the individual data themes that were originally entered into a given spatial database can be manipulated in combination to realise new information which was not present (at least not *explicitly* present) in the original.

One of the commonest forms of this is the manipulation of digital elevation models (see Chapter 5) to produce data themes which represent the slope or aspect of terrain. In this case the information about absolute elevation that is stored in the digital elevation model is used to realise potential information that would otherwise not be available. We will see that there are many other occasions when information that is implicit in a spatial database can be made explicit by manipulations.

There are a wide variety of ways in which GIS permits the derivation of new information but, before embarking upon a detailed discussion of these various options, it is worth making a brief point about the terminology we are about to use. In Chapter 1 we stressed that the thematic layers held within the spatial database are not strictly 'maps' in the sense of complex collections of multiply themed spatial data. However, when discussing the analytical capabilities of GIS packages the term 'map' has become synonymous within the GIS literature for thematic layer. As a result, rather than attempt to buck this trend you will find us freely interchanging between the words 'map' and 'thematic layer'.

4.1 THIS IS WHERE THE FUN STARTS

As mentioned above, there are a variety of ways in which the GIS allows us to derive new information. These can be divided into the following broad categories.

Searches or queries

These are operations that return a subset of the spatial database that meets some user-specified spatial, attribute or combined criteria. Although these are among the simplest type of operations possible, searching on spatial terms is one of the most widely used and valuable abilities of a GIS.

Summaries

Summaries can be obtained where one or more data themes are used as input, and tabular or other information is produced. Examples of this might be: obtaining the average elevation in a region, or counting the number of archaeological sites that occur on particular soil types. The first example requires only one input (a DEM), and produces a single figure as output (the average elevation) while the second would require at least two input data themes (site locations and soil zones), and would produce a table as output (sites listed against soil types).

Modelling techniques

These are methods for combining more than one map to produce a single map product. They are often referred to in the literature as *Map Algebra* or *Mapematics* for reasons that will soon become clear. Most simulations or spatial models require that several inputs are used to generate a resulting map. For example, a map of the estimated yield of a region could be calculated by performing map algebra operations on input themes such as crop, soil fertility and climate. An archaeological situation in which such map algebra is used is the generation of *predictive models* of archaeological site location. Here input themes (such as soil type, elevation, proximity to water) are correlated with the presence or absence of archaeological material in a pre-defined study zone and then combined according to the rules of the model to produce an output. This is then used to predict the location of archaeological material in areas beyond the original sample zone (see Chapter 8).

For alternative classifications of spatial operators, the reader might wish to review Burrough and McDonnell's approach (Burrough and McDonnell 1998:162-182) or Albrecht's (1996) classification which is very well discussed and illustrated by Wadsworth and Treweek (1999).

4.2 SEARCHING THE SPATIAL DATABASE

The ability to search the spatial database on geographic criteria, attribute criteria or a combination of the two, represents one of the simplest, but at the same time most powerful aspects of a GIS. In archaeology, this is useful both in cultural resource management situations and for research purposes.

Searching a database on the basis of attribute criteria is a routine archaeological task. For example, identifying all of the Prehistoric (chronological attribute) funerary monuments (classificatory attribute) from a regional monuments inventory. However, in management situations, an archaeologist frequently needs to query the database on the basis of *geographic* criteria *e.g.*:

- list the sites and findspots that occur in a proposed road development;
- list all the known sites that occur on commercially viable gravel deposits;
- list all of the sites that occur within the boundary of a National Park;
- find all the sites within 500m of a proposed pipeline cutting;
- list all legally protected sites in a 10km radius of central London.

Frequently researchers also need to query their data on the basis of spatial criteria in order to ask archaeologically meaningful questions.

- How many prehistoric sites occur within 2.5km of a source of flint?
- What proportion of Roman Villa sites occur within 5km of Roman roads?
- What proportion of Iron Age settlement sites occur within the viewshed of an Iron-Age hillfort?
- What are the proportions of locally made versus non locally made pottery present within 10km, 100km and 1000km of a kiln site?

These examples serve to illustrate the wide range of possible geographic (rather than attribute) queries that archaeologists need to express although they are obviously vastly simplified. In reality, searches often need to be phrased in terms of both attributes and spatial criteria and can be far more complex than the simple examples presented here. Fortunately GIS packages are adept at handling combined spatial and attribute queries.

Region searches for points

Region searches are queries that return data that falls within a specific geographic region. At the simplest level, this might be a query that returns all the archaeological sites that occur within a range of geographic co-ordinates such as *'list all sites whose x co-ordinate is between 1000 and 3000 and whose y co-ordinate is between 5000 and 6000'* (see Figure 4.1).

In fact this is such a simple query to formulate, that it would be straightforward to execute within a traditional (i.e. non-spatial) database, so long as that database stored the co-ordinates of the sites. In Structured Query Language, for example, it might be expressed as follows:

```
SELECT * FROM SITES
WHERE (SITES.X_COORD > 1000) AND (SITES.X_COORD < 3000)
   AND (SITES.Y_COORD > 5000) AND (SITES.Y_COORD < 6000)
```

Note that it is necessary to decide whether points that occur exactly on the region boundary are deemed to be inside or outside the region. In the SQL statement above, they would not be returned: to change this, we would have to change all of the 'greater than' and 'less than' signs (> and <) to 'greater than or equal to' and 'less than or equal to' (>= and <=) signs.

Using GIS the process is invariably much more straightforward. The simplest method is to specify the search region directly on the screen using familiar interface tools rather than type the co-ordinates as attributes. For example, by interactively dragging a 'rubber band' box over a given thematic layer.

The obvious limitations of this kind of search are that it is difficult to specify the search area precisely—there may be instances where we want to specify the exact co-ordinate dimensions of the selection box—and the fact that we are constrained to select a region that is rectangular and whose sides are parallel to the

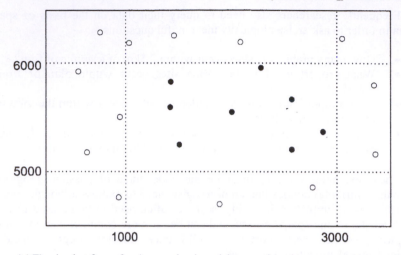

Figure 4.1 The simplest form of region search using minimum and maximum geographic co-ordinates to specify a geographic region as a rectangle.

X and Y axes of the geographic grid (see Figure 4.1). Far more commonly, we need to extract a subset of data that fall within an irregular region. Some GIS packages, for example ArcView, enable the user to interactively draw complex, irregular shapes as search areas, but once again this involves on-screen drawing and fails to overcome the problem of precise definition.

The routine solution to this problem is to specify the search area as a separate georeferenced polygon (see Figure 4.2). The GIS provides facilities for undertaking what are termed *point-in-polygon* searches on the two resultant layers, the theme you wish to search and the theme containing the search area. The precise manner in

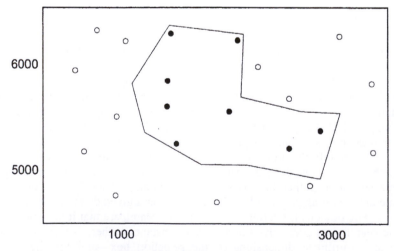

Figure 4.2 A point-in-polygon search.

which this kind of search needs to be expressed will depend on the particular GIS implementation—in raster GIS, for example, this is often undertaken as a specific case of a MASK operation (see below).

Region searches for lines and areas

Searching geographically for points is fairly straightforward—the only potential problem being points that fall exactly on the boundary of the search region. Do we treat these points as being inside or outside the search area? If we wish to search for higher order entities such as lines and areas, then life becomes more complicated because these have spatial extent (are 1- and 2-dimensional respectively) and as a result may cross the boundaries of the search region.

When dealing with line and area features, searches must be carefully specified. This is because a variety of different situations may occur. Figure 4.3 illustrates three criteria that might be used to select the polygons.

Firstly, we may set a criterion that asks for only those entities that fall entirely *within* the search region, as is the case for site A. So far so good. But what about features such as B and C that fall partly inside and partly outside the search region? These can only be selected by carefully phrasing the search to include sites that cross the region boundary.

Convex features, such as the semi-circular ditch D present a further problem, as it would not be returned based on either of these criterion despite the fact that we might well argue that the search region is likely to impinge archaeologically on the area enclosed by the ditch. To return site D in a search, we would need to search on the basis of either the bounding box of the polygon (shown as a dashed line) or its convex hull (shown as a dotted line). In most GIS systems this would require at least two stages of processing to undertake, in the first instance specifying the

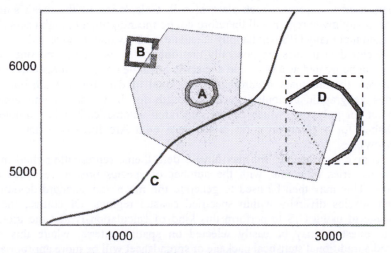

Figure 4.3 The relationship of the search region to three polygon entities is straightforward only for the ring ditch A.

required polygon and in the second undertaking the query.

4.3 SUMMARIES

Summary information about single map themes can take the form of either numeric summaries (statistics) or of graphical summaries, such as histograms or charts. Some of the summary operations, which may be present in any given GIS system, are described here, together with their application to some archaeological problems.

Of course, different GIS systems often use different names to describe their operators, although a user who is familiar with one group of notations will rapidly become familiar with a new set of descriptors. In general, the basic functionality which is described here will be available on all of the vector and raster GIS which the reader is likely to encounter, and it should be relatively straightforward to identify which software-specific command or menu option relates to which operator.

Univariate statistical summaries

As with non-spatial data, it may be important to gain numeric summaries of spatial variables to allow comparison between different regions or cases, or to quantify the variation or characteristics of some data theme. Most GIS will allow the production of values such as the minimum and maximum values present, the mean value and the standard deviation.

The precise operation that will be undertaken will depend on the data type that is being processed: some statistics will not be meaningful when applied to certain types of data. For example, much geographic information is in the form of categorical data such as soil type, which may be stored numerically, with different numbers used to represent each class of soil. Calculation of the mean soil type present in any given region will therefore not be meaningful, although users should be warned that many GIS will happily calculate such a value if requested.

Vector data themes are generally unproblematic, as each point, line or area feature can be treated as a case, and a specific attribute of the features may be used to generate the summary according to the level of data for that attribute. This includes geographic attributes, such as length for line features and area and perimeter for polygons, which can be generated automatically when topology is established for a given layer (as is the case with Arc Info) or when required (ArcView).

To give an example, we may have a data theme representing shell midden sites as a series of points, with the number of species present recorded as an attribute. This may then be used to generate the mean and standard deviation of midden species diversity within specified coastal regions. Of course, the only advantage of using GIS to perform this kind of calculation is that the groups of sites in question may be easily selected on spatial criteria; where this is not required a traditional statistical package or spreadsheet will be more appropriate.

Calculation of univariate summaries of choropleth data themes will also generally be straightforward to obtain in a raster system, where each raster cell will be treated as a discrete observation for the purposes of calculating such statistics. In undertaking such a summary it is important to remember that raster cells are essentially samples from an unknown population. As a result, the interpretation of such statistics is not without its problems. As already mentioned, the willingness of software to calculate a statistic is no guarantee that the statistic has any meaning or sense to it.

Area

One of the most useful fundamental GIS operators allows the production of summary area information from thematic layers. This is most commonly performed as an initial step in the analysis of spatial distributions, particularly where the relative proportions of some background variable are an important consideration. Area operations are generally performed on choropleth themes, such as geological or soil series maps.

Where the input data is in the form of continuous data, for example continuously changing slope values across an area, it is necessary to reclassify the input data into appropriate categories before calculating the areas in each class. In the case of the slope example given above we may want to reclassify the various degrees of slope into *flat, moderate, steep, vertical*. If this is not undertaken, then the results will vary from system to system: some will undertake the calculation regardless, and produce a table with a very large number of rows, others will deem it to be an error, and not perform the calculation.

In raster GIS the system obtains the value by counting the number of cells of each value, then multiplying by the area of each unit. In vector systems, the calculation is performed on the basis of the geometry of the polygons. As mentioned earlier, the respective areas of polygon features within a given thematic layer can be calculated automatically upon the establishment of topology or on a more ad-hoc basis, as and when needed. Because there is more calculation involved in the generation of area information in vector systems, raster systems will generally be faster at area operations. It is worth noting that as we are effectively 'counting cells' they do suffer from quantization errors, which will be greater with lower resolutions. Some vector systems maintain a record of summary information such as polygon areas, which means that the calculation need only be done when the geometry of the file is changed: in this case the calculation will seem immediate to the end user, although there will be a corresponding increase in the apparent time it takes to digitise or build topology.

The utility of an area operator can be illustrated with a simple example. We

Table 4.1 Hypothetical distribution of some archaeological sites on different soils.

Soil type	Site count	Soil type	Area
Soil type 1	2	Soil type 1	2.0 Hectares
Soil type 2	13	Soil type 2	153.0 Hectares
Soil type 3	4	Soil type 3	0.5 Hectares
Soil type 4	1	Soil type 4	44.5 Hectares

may have a spatial data theme that represents a region that can be divided into 4 categories of soil type. We may be interested in the distribution of some archaeological material across these soil types, observing that the archaeological sites in the region are not distributed evenly between the different soil categories (Table 4.1, left). From this, we might form an initial hypothesis that the settlements were preferentially distributed on soils of type 2. The first step in testing this hypothesis would then be to generate an area table from the soil map—clearly if there is much more of soil type 2 in the area, then we would be unable to support the hypothesis. Running an area operation on the soil map described above might produce Table 4.1 (right).

As suspected, there is indeed far more soil of type 2 in the region. However, we can also see that soil type 3 is barely present at all and that soil type 4 is extensively represented. With this information, we may modify our hypothesis to the extent that soil type 4 seems to have fewer settlement sites than we may expect, while soil type 3 may have rather more. With the information that an area operator produces, it is possible to be far more precise about the distribution of archaeological sites across spatial categories through the use of statistical hypothesis testing. In this case, it would be possible to construct a simple chi-squared test (Shennan 1997:104-126) with the null hypothesis that the sites are randomly distributed between the soil categories.

Perimeter

Another summary operator that is occasionally used produces a perimeter value for one or more classes of an input layer. As with the area operator, it will be necessary to classify continuous variables (such as elevation) into discrete classes before using this operator. This is similar to an area operation, except that the values in the resulting table represent a distance measurement for the length of the boundary that surrounds each class in the original map. This can be a useful measure of the variability of a map: two maps with the same number of classes may produce

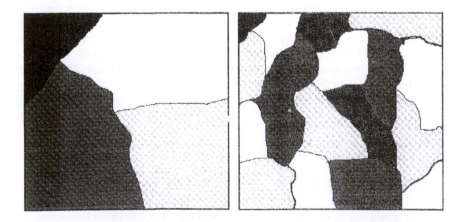

Figure 4.4 Area and perimeter statistics: both maps have the same number of classes, and produce the same area table but the map on the left has a very different perimeter table to that on the right.

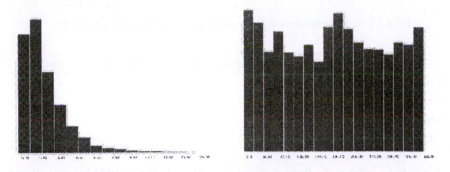

Figure 4.5 Image histograms for a slope map (left) and an aspect map (right).

considerably different perimeter tables if one is far more homogenous than the other (see Figure 4.4).

Unlike area calculations, it is easier to calculate perimeter values if the data is in vector format, when the system need only retrieve each vertex in turn and keep a running total. Perimeter calculations in raster systems are more difficult, because the function must allow for the fact that the raster cells are squares. Simply counting the raster cells that are adjacent to cells with a different value will not necessarily be a good estimate of the perimeter of an area.

Histograms and other graphical summaries

Summary information need not be presented in the form of a table. For example, if a map contains ordinal or higher level data, then the areas of each class can be presented in the form of a frequency histogram. This can produce an extremely valuable source of information, particularly as a first step towards investigating spatial relationships with statistical methods.

A frequency histogram will display the area of each class in the input map as a collection of graphical bars whose areas are proportional to the area of the class they represent. The form of the histogram can be highly revealing about the distribution of the spatial variable. As for area and perimeter calculations, continuous variables need to be reclassified into discrete classes before a histogram can be calculated, and users of GIS should be aware that the choice of interval or class size can have considerable effects on the form of the histogram which is produced.

The two image histograms in Figure 4.5 are frequency histograms for a slope map (reclassified into 15 classes) and an aspect map. The different character of each can be clearly seen. The slope map produces a histogram with a large number of low values and an exponential decline in the higher values. The aspect histogram shows that the area has a high proportion of north and south facing areas (the peaks in the middle and at the edges) but a lower area of east and west facing areas.

The frequency distribution of the slope values, left, is typical of many spatial variables, with a large number of low values and a decreasing number of observations as the variable increases. The distribution of the aspect variable is slightly problematic, because strictly this is not an ordinal variable in the same way

Table 4.2 The hypothetical result of a crosstabulation between geology and soil type.

	Rendsina	Grey rendsina	Brownearth	Gley	Podsol
Limestone	150	130	15	1	4
Clay	10	12	205	78	28
Shale	0	0	130	28	54
Alluvium	0.5	0	55	104	1

as slope but a polar one—there is no top or bottom of the variable because 0° and 360° represent the same value.

Crosstabulation

Given two maps, it may be important to know to what extent the classes of one map co-vary with the classes of a second map. The GIS operation required to undertake this process is referred to as the crosstabulation of two themes. For example, we may have a map of soil classes, and a map of geological types. If there are five soil types and four geological types we may crosstabulate them to obtain a table as follows. This indicates the area of each soil that occurs in each geological type. We can see immediately from this table that Rendsina soils tend to occur in areas of limestone, while Brownearths seem to occur in areas of clay or shale. We can also note that Gley soils seem to be associated with alluvial deposits (see Table 4.2).

Again, we should be careful that we do not simply assume that the numbers are significant: associations between the classes in two maps may be the result of an underlying cause, or they may be the result of chance. Instead, the output from crosstabulations can form the starting point for statistical significance testing, which will allow us to be rigorous in the ways in which we assess the covariation (see Chapter 6).

4.4 SIMPLE TRANSFORMATIONS OF A SINGLE DATA THEME

Reclassification and generalisation

Reclassification refers to an operation in which the attribute values of a choropleth data theme are changed according to some rule. It is one of the most basic of all map operations and often forms a prerequisite for other forms of analysis. It can be done in a variety of ways either from a legend, or automatically by specifying the class width or number of classes.

Reclassification is often a useful first step in modelling, or examining spatial relationships. It can also be used to convert between abstract measures and measures which give some specific meaning. For example, if we have a map of soil type expressed in terms of the soil series name, with four classes, we may wish to reclassify this into a new map which represents the *quality* of the soil for a particular purpose—say for crop production as good=3, medium=2, poor=1. The classifier would then be in the form of a *table* that related the soil type code to a soil quality code .

This table can then be used to reclassify the soil type map into a new soil quality map that may be of far more value to us as archaeologists in our investigations. Obviously the new map cannot have more classes than the original map, and in this case there is a generalisation from 4 to 3 classes involved in the operation. An example of precisely this type of operation can be seen in Gaffney and Stancic's study of the Dalmatian island of Hvar. Here 25 soil groups classified on the basis of factors such as chemical properties, physical characteristics and depth, were reclassified into 4 classes according to agricultural potential—very poor, poor, good, very good (Gaffney and Stancic 1991: 37-38).

Where the classes in a map can be grouped together, it is common to undertake a reclassification from a detailed map into a more general map. This represents a specific case of reclassification that is usually referred to as generalisation.

One characteristic of the reclassification process is that it results in the

Table 4.3 Hypothetical soil quality classification.

Soil type	Soil quality
1	2
2	1
3	3
4	1

generation of a series of classes into which the original input data must be allocated. The set of alternatives offered is clearly defined and unambiguous, what we can term a 'crisp' set (Eastman 1997, 9.24). Let us illustrate this with an example. We may have a thematic layer summarising the continuously changing density of ploughsoil material recorded over a study area. We may then decide that when a certain density threshold is exceeded in any portion of the surveyed landscape, this corresponds to a discrete activity area or site as opposed to a background manuring or discard spread. On this basis we can reclassify the original layer into site and non-site classes. The resultant map has no room for ambiguity—a location either corresponds to a site or it does not.

However, within much archaeological work there are considerable levels of uncertainty and ambiguity within the classificatory tasks we might wish to undertake. To return to the example given above, we may be dealing with a gradual increase in artefact density rather than a clear 'jump' between site and off-site making our classification highly artificial. In addition, if our reclassification threshold were set at 4 artefacts per metre square then any location with a density of 3.999 m² would automatically be classed as off-site whilst 4.001 would be regarded unambiguously as a site.

Fortunately, we can incorporate this uncertainty through the application of what is termed 'fuzzy set theory'. Fuzzy set theory comprises a number of concepts and techniques for handling the uncertainties that are associated in assigning an object to membership of a set of alternatives, where the transition from membership to non-membership is not abrupt but gradual (Zadeh 1980; Openshaw and Openshaw 1997:268-280). In his discussion as to the applicability of fuzzy set theory to real-world problems, Jain cites Zadeh's principle of incompatibility as the motivation behind the concept, where "In general complexity and precision bear an inverse relation to one another in the sense that, as the complexity of a problem increases, the possibility of analysing it in precise terms diminishes." (Zadeh quoted in Jain 1980:129). Jain considers Fuzzy Set theory to be a useful tool in complex situations where some variables are inherently ill-defined.

The incorporation of fuzziness within GIS-based approaches is a rapidly developing research field (for a seminal discussion see Fisher 1991, 1992, 1994) with some GIS packages, such as Idrisi, incorporating powerful fuzzy reclassification tools. In addition, a number of archaeological applications of such approaches exist, for example Gillings' work on the recreation of palaeo-flood impacts (Gillings 1998).

Mathematical transformations

Mathematical transformation operators are those in which a specific calculation is performed on all the geographic locations (features) within a given theme or set of themes. As an aside, this is a very similar definition to that which may be offered for the 'point processing' operations carried out in image processing. In each case the basic principles are exactly the same.

The simplest forms of mathematical transformation are scalar operations, which consist of the application of a single arithmetic operator with a constant value. Examples of scalar operators are addition, subtraction, multiplication, or division of a constant number applied to all areas in a spatial data theme.

To give a crude example, after a programme of intensive surface survey for a study area, we may have a layer in our spatial database which summarises the number of settlement sites per kilometre for a particular chronological period. If we know from excavation of a number of similar sites in the region that the typical

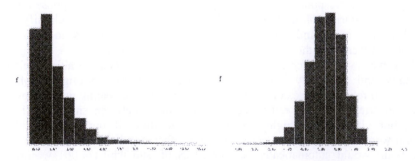

Figure 4.6 The frequency distribution of a slope raster (left) is transformed by taking the natural log of each cell. The distribution of *ln*(slope) (right) is a far closer approximation of a normal distribution.

settlement comprises 3 small dwellings, each home to a maximum of 6 people, it is possible to multiply the settlement density map by 18 to obtain an estimate of the maximum population density.

More sophisticated transformations may be used to change the distribution of data, for example to manufacture a spatial variable which approximates to a normal distribution from one which is not normally distributed. If the slope map from Figure 4.6 were transformed with a log operator, the distribution of the new map (which now shows 'log slope') appears closer to a normal distribution. The new transformed map may then be appropriate for use in certain parametric statistics, for which the original variable would have violated the normality assumption.

Other operations that come into this category include square, square root, trigonometric functions (sin/cos/tan *etc.*) and exponent operations.

Neighbourhood operators

Another class of operators provided by many GIS comprises those transformations that take account of whole areas of an input map, rather than simply calculating new values based on individual point locations. These kinds of context-contingent 'neighbourhood' transformations of spatial data are also important in digital image processing, where this kind of operation is termed a convolution (see *e.g.* Sonka *et al.* 1993), and find frequent application in fields such as the processing of geophysical survey data.

Like the class of operators discussed above, neighbourhood operators calculate the value of each location in the result based on some transformation of the information in the original. Unlike simple transformations, rather than work solely on the basis of individual point locations, these take account of a region around the point of interest in order to calculate a given result. In a raster system, this is achieved by considering a window of raster cells each time a calculation is made. What constitutes 'neighbouring' raster cells also varies according to the particular characteristics of the operator: at its simplest it means the eight immediate neighbours of a raster cell but it may involve more distant cells, and the window of cells chosen need not necessarily be square.

A mean operator

The simplest neighbourhood operators manipulate 3 x 3 blocks of raster cells, determining the result for each cell according to some combination of the nine values in the input. One of the most widely used of these, particularly in image processing, is a mean filter which serves to suppress 'noise' or random errors in continuous data.

A simple mean filter operation uses a 3 x 3 window of cells which is progressively moved over the input data theme. The mean of the value in all nine cells is used as the value for the central location in the new image. Variations on this include 5 x 5 mean filters, where the mean of 25 cells is used (or 7 x 7, 9 x 9 and so on). Figure 4.7 shows a simple mean filter in action. The values for the edge pixels of the new data theme are not shown. This is because they lack the requisite 8 neighbours for the calculation to take place. This can be overcome in several

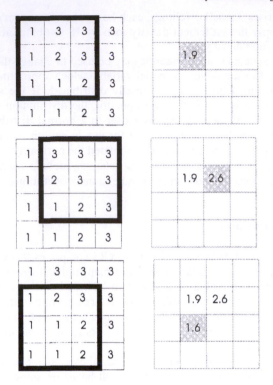

Figure 4.7 A simple 3 x 3 mean filter in action.

ways, depending on the specific implementation of the filtering algorithm utilised. The edge cells may not be processed at all, but simply copied from the input theme. Alternatively, the size of the template could be reduced to encompass only the actual neighbours of the pixel (so that the four corner cells would be generated from the mean of only 4 pixels, and the other edge cells would be the mean of 6). Clearly these will cause different results in the output data, referred to as *edge effects* (discussed further in Chapter 6). Edge effects are a characteristic of neighbourhood operators and should always be borne in mind when carrying out such operations.

Other examples of neighbourhood operators

By changing the rules by which new values are calculated, different results can be obtained. In the mean filter described above, the mean (defined as the sum of the observations/number of observations) was used to determine the values of the new data theme. Other mathematical properties may also be used, however, such as the maximum value in the window, or the minimum value. Some useful neighbourhood operators include:

- **Mode.** The mode of a set of values is simply the most commonly occurring value. Mode filters are often used to remove noise from a picture because they remove isolated abnormal values, replacing them with the local mode.

- **Median.** The median is an alternative measure of central tendency to the mean, it is defined as the value in a set of values (placed in rank order) such that half the values are above it, and half below it. Its effects are similar to that of the mean filter in that the image will tend to be smoothed, with isolated high values lowered and low values raised, but is a particularly useful filter in integer images because no rounding is necessary.

- **Variance and diversity indices.** These are operators which change the value of the target to some measure of the variation of the values present within the input window. For example, the variance is defined as the maximum value minus the minimum value (useful for ordinal data or above), while diversity might be defined as the number of different classes in the window (for nominal level data such as soil type). If we were to take a map of classes representing different environmental habitats, it may be interesting to calculate the diversity of habitats which are within a 1km radius of each location in the region. This might, in turn, be a good indication of the variety of resources which would be available for human communities to exploit. Diversity and diversity indices are discussed by various authors in Leonard and Jones (1989).

Examples of transformations

To give an example of the application of a neighbourhood transformation let us look at a specific problem that arose during the survey and study of a deserted medieval village in central Greece. One of the key research questions relating to the site concerned the location and form of the numerous house structures that detailed archival research had indicated were present at the site. As a result of post abandonment robbing for the probable construction of sheepfolds and field boundaries, compounded by a low, but dense, mat of prickly oak that obscured large areas of the site environs, on initial inspection domestic structural remains appeared to be non-existent (see also the example in Chapter 5).

In an attempt to capture any surviving traces of the habitation areas surviving beneath this dense carpet of vegetation a detailed micro-topographic survey was undertaken using a total station to log a regular grid of elevation points across the area. Once the data had been collected and collated, the GIS package Arc Info was used to construct a Digital Elevation Model, *i.e.* a continuous representation of the surface topography of the area. We will be discussing DEMs in detail in Chapter 5. A number of interesting areas of particularly undulating topography were immediately apparent and in an attempt to further investigate by highlighting any rectangular house platforms that may be present a 3 x 3 edge enhancement filter was applied to the raster DEM. This is referred to in the Arc Info literature as a 'high pass' filter. In practice the 9 values in the 3 x 3 window were weighted as follows:

```
-0.7   -1.0   -0.7
-1.0    6.8   -1.0
-0.7   -1.0   -0.7
```

The result is a raster layer where each cell value represents the smoothness of the surface at that point (ESRI 1997). In practice the filter served to highlight a series of areas where rectangular house platforms could clearly be identified (Gillings 2000).

4.5 SPATIAL DATA MODELLING

Transformations which involve interactions between more than one data theme (rather than between one data overlay and a constant as above) are sometimes referred to as 'map algebra' techniques or 'mapematics' (Tomlin 1983, 1990). All GIS, both raster and vector based, should provide a mechanism for combining different data themes using either arithmetic or logical (boolean) operators as this is the basic mechanism of spatial modelling.

Arithmetic between layers

Most GIS will provide facilities to add, subtract, multiply or divide one map from another map. In these operations the attribute value of each location in the new map is calculated from the values present at the same location in the source maps. For example, if we have a map showing the density of prehistoric sites in a region and a second map showing the density of historic sites and want to know the overall site density we would proceed by adding the two maps together in this way.

It is important to realise that within a GIS, the full range of mathematical

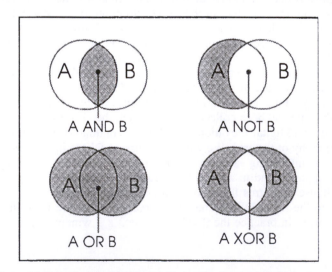

Figure 4.8 Venn diagram showing the effects of the four basic boolean operators.

operators which can be used with a constant can also be used with another map. As a result maps can be subtracted from one another, divided, multiplied, one can be raised to the power of the other and so on. One application area within archaeology that makes extensive use of these capabilities is in the field of predictive modelling. For example, in generating predictive models for the location of prehistoric sites in the Upper Chesapeake Bay area of the US, Westcott and Kuiper (2000) used map algebra to combine four input layers (elevation; distance to water; quality of water; topographic setting) into a single layer encoding potential for site location. Predictive models will be discussed in detail in Chapter 8, and we will discuss the Chesapeake Bay area study in more detail there.

Boolean operations

As well as these arithmetic operations, it is possible to define a set of logical operations between maps. These are expressed with the boolean operators AND, OR, NOT and XOR. The effects of boolean operators can be summarised in the Venn diagram shown in Figure 4.8. Boolean operators return either TRUE (usually coded as a 1) or FALSE (coded as a 0). As a result, a map generated by such a process will always contain only those two classes.

Boolean operations are of particular value in unique conditions modelling. For example, as cultural resource managers, we may be interested in identifying zones which are particularly prone to soil erosion within our area of responsibility. Our input data layers comprise a map of soil depth (in 3 classes ranging from 1= thin to 3 = thick) and a map of slope (in 5 classes, where 0 = flat to 5 = very steep). If we assume that a combination of steep slopes and thin soils result in erosion then we can use the following boolean map operation to identify our zones:

{ (SLOPE=4) OR (SLOPE=5) } AND (SOIL=1)

Note that the additional brackets are necessary in this expression because unlike arithmetic operations, boolean expressions are not *commutative*. For example: the expression (A + B) - C gives the same result as A + (B - C) but (A OR B) AND C will not necessarily give the same result as A OR (B AND C).

Mask and cover overlay operations

Although they are officially specific cases of boolean operation, two of the most useful operations which can be performed between two data layers are sometimes considered as entirely separate operations. These are termed 'cover' and 'mask' overlay operations. If A and B are maps then:

• **Cover A with B** will produce A where B is 0, otherwise B. In effect map B is treated as if it were drawn upon a glass sheet - where no features are recorded the features present on the underlying map sheet A can be seen. In boolean terms this is IF (B=0) THEN A ELSE B.

- **Mask A with B,** where B is usually a binary image, is a particular case of a cover operation. It will produce A where B is non-zero and zero otherwise. This can be expressed in boolean terms as IF (B=0) THEN 0 ELSE A.

To illustrate the usefulness of such operations let us imagine that we are interested in undertaking a classic site-catchment analysis of the type originally proposed by Vita-Finzi and Higgs (1970). Our aim might be to quantify the types of landuse present within a catchment circle centred upon a settlement site, in an attempt to shed light upon its subsistence base. A first stage would be to create a mask layer representing the catchment circle. This might be a vector polygon or a raster layer with a value of '1' given to all cells falling within the catchment zone and '0' to those outside it. We could then use a mask operation to overlay this upon a layer encoding landuse zones. In effect the mask layer acts as a cookie-cutter, extracting just those portions of the respective landuse classes present which fall within the catchment zone. It is then a routine task to summarise the areas of the landuse classes present in the catchment circle in an attempt to characterise the subsistence base. We will discuss approaches such as site catchment analysis in more detail in Chapter 7.

CHAPTER FIVE

Digital elevation models

"Pooh! Stuff! Enough of this trifling! The time is short, and much remains to be done before you are fit to proclaim the Gospel of Three Dimensions to your blind benighted countrymen in Flatland." (Abbott 1932: 87)

The previous three chapters have deliberately avoided discussing elevation data and this has been for very good reason—it is so widely used as a primary source of data that it really requires a separate chapter to itself. Because it is a model of continuous variation, rather than discrete classified data, Elevation presents some particular challenges for both its storage and manipulation. In a sense, therefore, this chapter is best thought of as adding a specific example to Chapters 2, 3 and 4 and its remit is to examine how we can use GIS to store, capture and manipulate this particular class of data.

The first thing to note is that elevation models differ fundamentally from the data we have discussed up to now. Geology maps, soil maps, political areas or vegetation types can be regarded as discrete variables, in that there are a finite number of categories into which the land surface is divided. An elevation map, on the other hand, represents continuous variation over the land surface. This is usually, but not always, a map of the topographic height of the terrain, normally expressed in metres above sea level.

5.1 USES OF ELEVATION MODELS

The storage and manipulation of terrain data is one of the tasks for which GIS are most frequently used. Elevation data—and its derivatives such as slope and aspect—have been used to underpin a very wide variety of analytical methods including:

- visualisation of topography and of other data in relation to terrain;
- cost-distance and least-cost pathway analyses (see Chapter 7);
- predictive modelling for research or management (see Chapter 8);
- analysis of visibility and intervisibility (see Chapter 10);
- simulation of natural processes such as flooding and erosion;
- virtual reality and the visual recreation of archaeological landscapes.

The term *digital elevation model*, commonly abbreviated to *DEM*, is the most widely used term for models of the earth's surface, although it is worth noting that it need not necessarily refer solely to surface topography or landform. In fact any empirically measurable variable that varies continuously over space can be regarded as an elevation model—pottery density, distance from water or rainfall for example. As such, when landform is being described the term *digital terrain model* might be preferable because this specifically identifies the data as a model of topographic height. Following this definition, all digital terrain models are examples of digital elevation models, but it is worth remembering that some digital elevation models may not be of terrain. In this chapter, we will be primarily concerned with models of landform and their particular characteristics.

5.2 ELEVATION DATA IN MAPS

On paper maps, terrain is usually represented by *contour lines,* which are simply lines connecting together points of the same topographic height. Contours give a reasonable visual representation of terrain form, but in many ways they are far from an ideal way of representing terrain. Variation which takes place between the contour intervals can only be fully represented by either adding supplementary contours, or by including additional representations of terrain such break-of-slope lines or spot heights.

As we have made a distinction between digital *elevation* models and digital *terrain* models, we should also make the observation that the term *contours* is usually restricted in use to representations of terrain, while the more general term *isolines* can be used to refer to the same notation used in other situations. For example, we may want to draw isolines to communicate changing densities of pottery within the ploughsoil of a survey region. There are also some special terms for particular types of isoline: isolines of barometric pressure are called *isobars,* and for temperature maps *isotherms.*

Contour lines are not the only notation for topography we find on maps. Occasionally other representations are used, such as coloured areas or shading although these usually involve a substantial generalisation of the topographic information and tend to be used for display purposes only. One method of representing topographic variation extensively used in archaeology, particularly field survey and in the production of drawings for publication, is the hachure plan. This can be used to indicate a surveyor's interpretation of very fine variations in topography, including scarps and changes of slope that may not be visible on contour maps unless a very fine contour interval is used. Hachure plans have a series of fixed conventions, so that closely spaced hachures represent steeper slope while greater spacing indicates a shallower slope. The shape of the hachure 'head' and the nature of the 'tail' are also used as indications of whether there is a gentle or marked break of slope. Hachure plans are a useful and well understood notation within archaeology and can be used to create clear visual representations of terrain form that would be difficult to represent using contours (see Figure 5.1).

However, hachures are purely qualitative in that they accurately represent only the relative variation of the terrain rather than absolute height (although clever use of spot height indications combined with hachures can represent a topographic surface reasonably comprehensively). They also tend to focus upon interpreted detail at the expense of broader topographic trends and hachure plans can be so densely drawn and complex that they fail to convey any coherent meaning whatsoever[4]. From a GIS perspective, hachure plans are impossible to

---

[4] The worst example of the use of hachures may be the plan of Grimes Graves, Norfolk, produced by the Royal Commission for Historic Monuments of England. This closely resembles abstract art, conveys no meaning whatsoever and at the same time provides no opportunities for quantitative analysis. High reproduction fees and unhelpfully restrictive licence prevented reproduction of that image here.

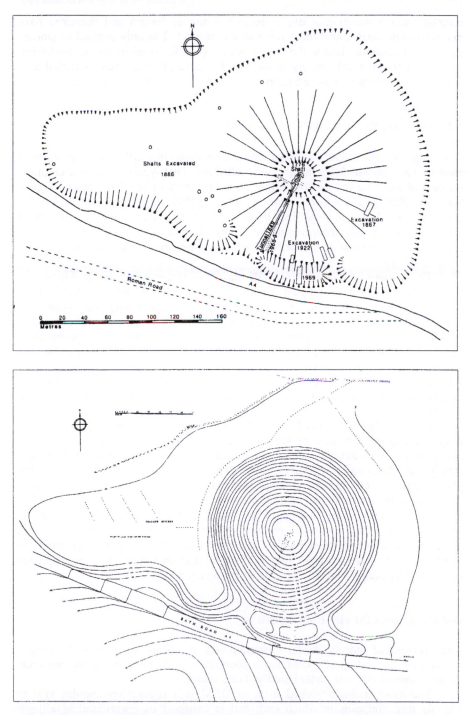

Figure 5.1 The site of Silbury Hill, Wiltshire, England represented as a hachure plan (top) and as a contour survey (bottom). From (Whittle *et al.* 1999), reproduced with permission.

translate into a meaningful digital representation of terrain and therefore have extremely low formal analytical value in this respect. The only method of storage suitable for hachure plans within a GIS is as a georeferenced image or 'backdrop' whose use is restricted entirely to display. Hachure plans are best regarded as a form of interpretation or presentation rather than as a formal method of recording.

5.3 STORING ELEVATION DATA IN GIS

The simplest form of elevation data that can be stored in a GIS is a series of point entities, each of which has height as an attribute. This type of data is a common product of field survey and can either be in the form of regularly spaced points, usually called *gridded point data*, or as irregularly spaced points which are generally called *spot heights*. Spot heights have the advantage that the density of observations can vary according to the degree of variability in the terrain and this can reduce the amount of data redundancy that inevitably occurs in flat areas of a gridded survey.

The storage of contours in a GIS database presents no new problems that have not already been discussed. Contour lines can simply be regarded as lines with an attribute corresponding to known height or nested polygons, again with height as a single attribute. Contour plans can therefore be stored in the same way as other types of choropleth data. In fact, contours present rather fewer problems in this regard than other overlays because contour lines never cross one another and never meet, which means that there are no topological links to make between them. Although contours are easy to store, they do not really constitute an elevation model *per se* because a contour map does not form a continuous model of the terrain.

In fact, while contours form a good visual representation of terrain that is easy to interpret—not least because we are so familiar with it—contour maps do not present a particularly good source of elevation data from a statistical standpoint. The contour interval can serve to mask or generalise variation that takes place between intervals unless 'supplementary' information, such as intermediate contours, scarp lines and spot heights, are also included.

Nevertheless, so much topographic data is in the form of contours on paper maps that digitised contours are commonly used as a first stage in the process of generating an elevation model and as an output format for representing continuous variation in printer or plotter output. Terrain data in the form of pre-digitised contours is available from a number of mapping agencies.

Data structures for elevation models

The problem of storing digital elevation data is essentially one of devising a discrete representation of a continuous phenomenon. As such, there are two quite distinct approaches, one raster based and one vector.

The most frequently used form of DEM is a regular rectangular grid of altitude measurements in which each cell is assigned its corresponding altitude value (Figure 5.2). Such a matrix can then be manipulated in exactly the same way as all other raster image files, such as satellite images or scanned maps. Many

Figure 5.2 Altitude matrix presented as a 3D wire mesh. Data represents part of the upper Kennet valley, S. England, generated with GRASS d.3d.

mapping agencies now supply topographic data in this format, including the United States Geological Survey and the Ordnance Survey of Great Britain.

Commonly, altitude matrices are generated automatically from point data or from digitised contours—a process that is a specific case of *spatial interpolation*. Archaeologists who consider purchasing elevation data would be well advised to carefully read the sections on interpolation (both this chapter and Chapter 9) and should expect data providers to give details of the source data and form of interpolation used to generate the data.

Altitude matrices are the most widely used form of elevation model for the generation of slope maps, aspect maps, analytical hill shading and for the analysis of visibility but there are some inherent problems with this type of model. Firstly, it is not always entirely clear what the value in each cell represents. It is probably best regarded as the average topographic height within that landscape unit, but could just as easily represent the maximum height, or the height at the centre of the unit. The latter will be the case, for example, where an altitude matrix is interpolated from gridded survey data with the same sampling resolution. This issue is even more pronounced when we consider the raster derivatives of elevation, such as slope and aspect where it has been demonstrated that the computed slope or aspect values best represent a cell size 1.6 to 2.0 times that of the original altitude matrix (Hodgson 1995).

Because the sampling interval of an altitude matrix is the same for areas of high variability as for very flat areas, there is usually considerable data redundancy in altitude matrices. This can lead to very large data files, and consequently to very slow processing of data. In fact, in common with all raster layers, the choice of resolution is critical when using elevation data. If the cell size is too small then the file will be very large, with massive redundancy, while if the cell size is too large, small scale variations which may be significant will be obscured. Given the ubiquity and inherent problems of altitude matrices, it is surprising that, with a few notable exceptions (Kvamme 1990b, Hageman and Bennett 2000), very little attention has been directed in the archaeological literature to their use and derivation.

An alternative to the altitude matrix is the vector-based Triangulated Irregular Network (TIN) model of elevation. This was devised in the late 1970s (Peucker *et al.* 1978) as a flexible and robust method of representing elevation that is more efficient than regular matrices whilst at the same time allowing for the calculation of valuable derivative maps such as slope and aspect.

A TIN consists of a sheet of connected triangular faces, usually produced from a Delaunay triangulation of irregularly spaced observation points. The Delaunay triangulation is discussed further in relation to spatial allocation problems (see Chapter 7), but for now it is enough to know that it is an 'optimal' method of choosing how to make a set of triangular faces from irregularly spaced point data (Figure 5.3). TINs can easily be built up from irregular '*x, y, z*' point

Figure 5.3 The site of El Gandul, Andalucia. The image is a 3D projection of a TIN with the shading of the faces derived from analytical hillshading. The site's defensive earthworks show clearly as an interrupted bank towards the left of the image. Note also the ridge-like structures running along the slopes—artefacts of the TINs origin as a contour map.

observations which may result from survey work. Unlike regular matrices, this allows TIN models to include higher densities of observations in areas where this is most important—areas of high variability—while using fewer observations for relatively flat areas. This allows the representation of complex and highly variable terrain without the need for a great deal of redundant data. Nodes of the TIN can also be arranged so that the model specifically follows ridges or valleys and respects flat areas such as lakes.

TIN models are stored as vectors, with the nodes storing height values. They can be regarded as a simple form of the arc-node-area data structure discussed in Chapter 2, with several simplifications: the line entities of the model always have only one vertex; all area entities are defined by exactly three lines; and there is no need to code islands and envelope information as these never occur.

Although altitude matrices and TINs are fundamentally different methods for storing elevation models, methods for translating between the two have been proposed and implemented in some systems (examples include Arc Info and ArcView 3d analyst). Translation from TIN to matrix is relatively straightforward, involving laying a grid over the triangulation and assigning the appropriate height value for each grid cell according to the value either at the centre, or calculated through some interpolation algorithm.

5.4 CREATING ELEVATION MODELS

The ideal method of obtaining accurate elevation data is probably to go out into the field and survey it using either conventional survey equipment—in other words theodolite or total station—or with GPS (see Chapter 3). Survey, however, can be impractical if the study area is large or inaccessible and can also be prohibitively expensive. As a result, most archaeological applications have turned to existing sources of terrain data. Until very recently this has meant topographic maps, usually published with contours and some spot height data. More recently, photogrammetrically-derived elevation data has become more widely available to archaeological projects and this has begun to be a viable alternative to the use of map-based elevation data.

Interpolation

Altitude matrices and TINs can be interpolated using a variety of methods, most of which will be discussed in Chapter 9, when we consider interpolation more generally. However, it is useful here to describe one special method of interpolation that is unique to the generation of altitude matrices: the interpolation of height values from contours. The importance of this stems mainly from its widespread—and usually rather uncritical—use as a source of data for archaeological analysis.

The most common source of elevation data is contour maps while the most frequently used form of elevation data is the altitude matrix. Because of this, altitude matrices interpolated from contour data probably represent the most widespread form of elevation model that has been used in archaeology. Translating

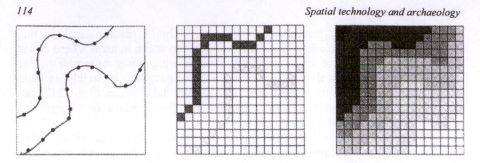

Figure 5.4 Possible stages in the creation of a DEM. Left: vector contours, middle: rasterised contours and right: interpolated elevation model.

between these two forms of elevation model is generally undertaken in three steps (shown graphically in Figure 5.4):

1. digitise the map contours into a vector file;
2. rasterise the vector contours;
3. interpolate the height values between the contour lines.

It is also possible to interpolate height values directly from vector line data, thus removing one source of the quantization errors from the process. For simplicity, though, we will explore the three-stage approach that is implemented in a variety of raster GIS (this particular example is based loosely on the GRASS *v.surf.contour* module).

The first two steps are relatively straightforward, although we might draw attention to the need for the software used to be consistent in whether it regards cells as being directly connected to 8 or 4 adjacent cells when forming lines. Once the contours have been translated into a raster form, the process of calculating the values for the other cells in the matrix can be undertaken. This can use a variety of different algorithms which each use a methods of calculating the values for the heights in between the contour lines.

The effect of different algorithms

Figure 5.5 shows a typical example of the task of an interpolation algorithm. After rasterising the contour lines (10, 20, 30 and 40m) the algorithm must ascertain values for all those cells not occurring on a line. The unknown cell is labelled *e* while eight possible sources of information about the unknown height are labelled 1 to 8. The elevations at each of these points can therefore be given the notation e_1 to e_8 while the distances from the point *e* to each point can be denoted d_1 to d_8 respectively. For points 1, 3, 5 and 7 these are easily obtained but for points 2, 4, 6 and 8 the Euclidean distance must be calculated using pythagoras' theorum.

Perhaps the simplest interpolation algorithms, but least reliable, are *horizontal scan* and *vertical scan* algorithms. These scan the cells of the matrix from left to right, or top to bottom, averaging out the values between known cells. If the elevations at the points 1 to 8 are given by e_1 to e_8 respectively, and the

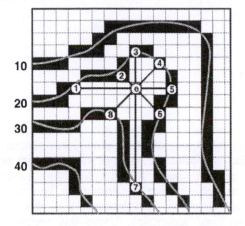

$$e_1, e_2, e_3, e_4, e_5, e_6 = 20, \quad e_7 \text{ and } e_8 = 30$$
$$d_1 = 5 \quad d_2 = 1.41 \quad d_3 = 3 \quad d_4 = 2.83 \quad d_5 = 3 \quad d_6 = 2.83 \quad d_7 = 7 \quad d_8 = 2.83$$
$$s_{1,5} = 0 \quad s_{2,6} = 0 \quad s_{3,7} = 1 \quad s_{4,8} = 1.79$$

Figure 5.5 Interpolating between contour lines.

distance between two points x and y is indicated by $d_{x,y}$ then the height of the unknown cell in Figure 5.5, by a horizontal scan algorithm is given by:

$$e = \left((e_5 - e_1)/d_{1,5} \right) + e_1 \tag{5.1}$$

The possible errors that this type of approach can produce can be deduced by looking at Figure 5.5. A horizontal scan algorithm will clearly base the new value only on the 20m contour because of the loop in the line. Using equation 5.1 the slope at the unknown cell will be calculated as 20m, which is clearly too low. More satisfactory are *steepest ascent* algorithms which search in either four or eight directions from the unknown cell, ensuring that the new value is based only on the contours which produce the steepest slope. The algorithm then uses the steepest obtained value as for the vertical and horizontal scan algorithm. If we use the notation $s_{x,y}$ to indicate the slope between points x and y then the steepest slope of the 8 directions can be specified as:

$$e = \max \lfloor s_{1,5}, s_{2,6}, s_{3,7}, s_{4,8} \rfloor \tag{5.2}$$

In our example, the steepest slope occurs between points 4 and 8 and the unknown value will be calculated to be 25m.

Another alternative to this is a *weighted average* algorithm, which uses the same values as above but instead of using only the steepest slope, replaces the unknown cell with the average slope value, weighted for the distance to the contour lines. This can be expressed as follows:

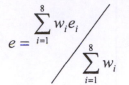

$$e = \frac{\sum\limits_{i=1}^{8} w_i e_i}{\sum\limits_{i=1}^{8} w_i} \tag{5.3}$$

The weights are designed so that the estimate of elevation for the unknown cell is biased towards the closest of the contour measurements, and a function needs to be chosen that approaches the value of a known point as the distance to it approaches zero. A popular choice for this weighting function value is the reciprocal of the Euclidean distance between each known elevation and the unknown location, and algorithms that interpolate in this way are known as 'inverse distance weighting' algorithms. Using this method, the value of e, calculated from Equation 5.3 will be 21.72m.

There are many more variations on these methods, and more optimal forms of interpolation have been devised, but what should be clear is that the algorithm employed to interpolate the elevation model can substantially affect the result. In our example, it is unclear which algorithm gives a better approximation of the actual elevation, although the simple scanning techniques seem intuitively to be less reliable. Kvamme (1990b) has experimented with a variety of such algorithms and demonstrated the significant differences that can result from the use of the different interpolation algorithms discussed here. Figure 5.6 summarises some of the results. Notice how the vertical scan algorithm produces a marked 'stepping' of the model, oriented in the direction of the scan. This is clearly an artefact of the algorithm, and the result would be unusable in many situations without further processing.

Vertical or horizontal scan algorithms can produce a marked stepping of the digital elevation model or other artefacts such as the appearance of a 'faceted' model, in which the contour lines appear as breaks of slope. This type of result is particularly noticeable if contours are widely spaced compared with the cell resolution, and some form of steepest ascent algorithm is used.

In many applications, such as the calculation of slope or aspect maps, these effects may produce unacceptably distorted results and to circumvent such problems it is not uncommon to smooth the elevation surface before it can be used for the generation of derivatives. This is generally done by using a mean filter, either a simple convolution filter or a weighted mean operator, which produces less drastic results (see Figure 4.7).

Although filtering an elevation surface can be necessary to improve the continuity of the surface, which can be beneficial for the derivation of slope or aspect derivatives, it is important to note that this process introduces yet more algorithmic-induced distortions to the elevation matrix. A filtered model may no longer show the abrupt changes of slope that can be produced by interpolation algorithms, but smoothing also affects the height of the surface at the contour lines. Figure 5.7 shows how a mean smoothed surface will tend to be lower than the contours on hillsides while higher than contours in valley bottoms.

Obviously this is an undesirable consequence of filtering the terrain but the loss of absolute accuracy must be balanced against the better representation of the terrain character that results from smoothing abrupt changes of slope. In situations where derived information depends on the landform more than the absolute height of the terrain (such as in the calculation of slope or aspect maps) then smoothing may be a necessary step. However, in other situations the absolute height of terrain may be of considerable value and the loss of information might be deemed

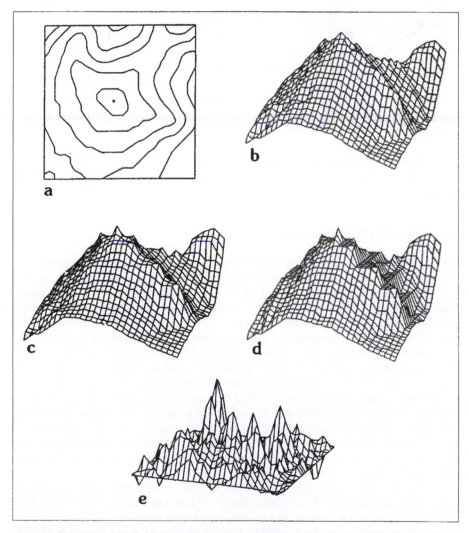

Figure 5.6 The effect of different interpolation algorithms. (a) Contour map and DEMs produced by (b) steepest slope algorithm, (c) weighted average algorithm, and (d) vertical scan algorithm. (e) The difference between b and c (Kvamme 1990b:115 Figure 10.2).

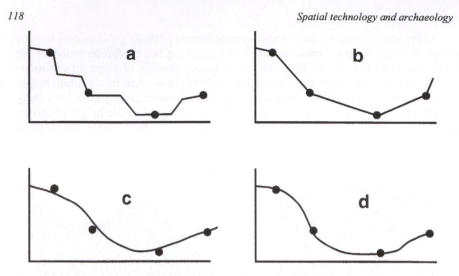

Figure 5.7 Cross sections through hypothetical elevation models. (a) Vertical scan, (b) steepest ascent, (c) typical effect of filtering, and (d) the real elevation values. The black dots represent the contours.

unacceptable. Either way, archaeologists who deal with elevation models would be well advised to avoid the use of post-processing in this way unless it is clearly justified.

Generating TINs from contours

Further considerations arise when vector contours are used as the primary source of elevation data for the generation of a TIN. Although contour data is perhaps the most commonly available source of elevation information it is also the most difficult to use in TIN creation (ESRI 1997). Firstly, to create the nodes of the TIN, contour lines have to be regularly sampled to yield a set of points. For example every 30 metres along the course of a contour line a point may be extracted. Most TIN creation algorithms (for example CREATETIN in Arc Info) enable you to specify this sample interval (through specifying *weed* and *proximal* tolerances). With relatively uniform terrain this may cause few problems. However, when there are complex topographic features *e.g.* spurs or promontories, there is a good chance that none of these resample points will adequately capture the complexity of the feature and highly significant topographic points are missed. In addition, given that in a TIN groups of 3 points are linked by a triangle, in the case of a looping contour these 3 points may be taken from the same contour line leading to an artificial 'flat' triangle in the TIN.

Solutions to these problems are firstly to ensure that any additional topographic data that may be available (for example spot-heights) is incorporated into the TIN generation process. In addition, rather than inputting contours as lines, you might instead undertake a process termed *intelligent point selection* (ESRI 1997). Here, instead of digitising contours into the spatial database, you instead manually digitise elevation points from the contours (Figure 5.8). In effect you are replacing the automatic (and undiscerning) resampling procedure interactively, ensuring that all critical points along the contours are included in the TIN generation procedure. It should be noted that whilst intelligent point selection results in much more robust TINs, to use it effectively it is important to understand how the triangulation process works.

Photogrammetry

We have already seen in Chapter 3 how digitised aerial photographs can be georectified to create *orthophotos* for inclusion in a GIS database. However, aerial photographs also provide a major source of elevation data through the application of *photogrammetry*. The term photogrammetry actually refers to a large field of study concerned with obtaining accurate measurements from photographs. Here we are interested in those photogrammetric methods that allow elevation to be calculated from stereo pairs of photographs. This relies on measuring image *parallax*, the apparent change in the relative positions of objects in the photographs when viewed from different places.

As with many other interesting areas, the constraints of space prohibit us from going into any real detail about photogrammetric methods, although the interested reader who wants to learn more about photogrammetry might start by reading Chapter 4 of Lillesand and Kiefer (1994). In the context of digital elevation models, we should note two general methods that can be used to produce elevation data.

The first is the use of stereoscopic plotting instruments to prepare topographic maps. These are essentially instruments that allow an operator to see a stereoscopic view derived from overlapping aerial photographs. The operator sees a floating mark, which can be raised and lowered as it is moved across the terrain so that it maintains 'contact' with the viewed surface. A connected stereoplotter or

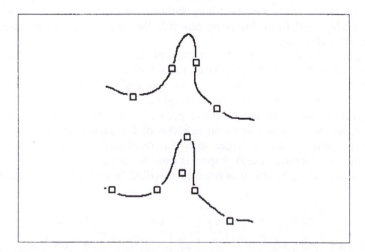

Figure 5.8 Intelligent point selection. Top: the contour line has been resampled automatically and sampled points do not best capture the topographic trend of the contour loop. Bottom: an intelligent point selection results in optimum points being selected, in this case an additional spot-height has been included to avoid a flat triangle occurring within the loop.

digitiser can be used to 'trace' out contour lines either onto paper or directly into a computer file. Although computers are often used with these instruments, this is essentially a manual method because it relies on an operator to place the mark and so create the data. It can also be a labour-intensive method of producing elevation models, although obviously far less so than ground survey.

The second area of photogrammetry is potentially far more exciting. This is the arena of *digital photogrammetry* in which the aerial photographs are scanned, and the processing is undertaken with minimum intervention by an operator. Digital systems can automatically output accurate elevation data in the form of contours, altitude matrices or Triangulated Irregular Networks.

5.5 PRODUCTS OF ELEVATION MODELS

Elevation itself can be analytically useful in some situations—we may be interested in the absolute elevation of archaeological sites, or we may wish to estimate or model a hypothetical flood or the location of the tree line, for example. However, elevation data can be used to generate a very large number of *secondary products* which can be even more useful to archaeological analysis.

Slope and aspect

Calculation of slope maps from altitude matrices is relatively straightforward, and directly analogous to the use of *gradient operators* in image processing. Slope is usually defined by a plane tangent to the surface of the elevation model and in reality consists of two components: *gradient*, which is the maximum rate of change of altitude within that plane and *aspect*, which is the direction in which the maximum rate of change occurs. Many authors use the terms *gradient* and *slope* interchangeably, and from this point onwards the term *slope* can be taken to be synonymous with *gradient*.

Slope can be measured either in degrees (from 0° to 90°) or in percent, in which case 0° is 0% but 90° is referred to as 100%. Slope is generally estimated from the geometry of the values in a local neighbourhood which may include four, eight or more values which may also be weighted (Figure 5.9).

Aspect is also estimated from the geometry of a local neighbourhood and, like slope, has been shown to be an estimate of the surface orientation of a cell between 1.6 and 2.0 times larger than the resolution of the elevation model (depending on the method used). Aspect is usually calculated in degrees with either 0 or 360 representing North. It is often reclassified from degrees (or radians) into

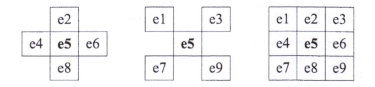

Figure 5.9 Possible neighbourhoods for estimation of slope and aspect values at location e5.

the eight main compass directions (N, NE, E, SE, S, SW, W and NW).

There are two things to note about aspect data. Firstly, we need to be aware that it is not readily amenable to statistical treatment, being a *polar* scale measurement rather than an interval or ratio scale one. Put simply, this means that we cannot treat it in the same way as, say, elevation or rainfall measurements because there it has no minimum or maximum values. Statistically this can make things complex.

Secondly, there should always be a separate 'no aspect' value which is specified whenever the calculated slope is zero, *i.e.* the surface is flat. This too can complicate the use of aspect themes in statistical analysis (such as regressions) because any inclusion of aspect is also incorporating some information about slope.

Analytical hill shading

One of the most familiar products of elevation models is the analytical shading map. This is closely related to slope and aspect, using both to estimate the intensity of reflected light that would be seen by an observer who was situated perpendicularly above the terrain. Given slope and aspect maps, the calculation requires only that the location of the light source be known. Most are made on the assumption that the reflectivity of the elevation surface is constant: in other words that the terrain is not more reflective in particular places. It follows from this that the intensity depends only on the relationship of the slope and aspect to the light source.

Results can be extremely convincing, if rather reminiscent of a desolate lunar landscape, and are especially effective when visualised 'draped' over a three-dimensional 'mesh' diagram of terrain. Such products are not only visually attractive, but can be very useful for the visual interpretation of landform, as they represent terrain in a way familiar to the eye, allowing subtle features of the terrain (and, it should be added, artefacts introduced by the interpolation procedures) to be rapidly picked out.

Hydrological and flood modelling

Another application of elevation models is to hydrological modelling. Once again, it is only right that we point out that hydrological modelling is itself a substantial area of study that our volume can only draw attention to.

Modules to undertake simple simulations of hydrological processes are available in several commercial GIS, and have obvious applications in archaeology. Most operate by simulating the direction and magnitude of flows in either an altitude matrix or TIN, and keeping a cumulated count of the flow of water into and out of each element of the model. The models can be simple or complex, with more complex models allowing some account to be made of the different characteristics of the terrain that will impact on the movement of water across or into the ground. Obvious applications of hydrological modelling within archaeology include the possibility of estimating the course of ancient rivers and stream networks, or identifiying those parts of archaeological sites which are most at risk of erosion.

MICRO-TOPOGRAPHIC SURVEY: GREEK MEDIEVAL VILLAGE VM4

A good illustration of the usefulness of products such as slope maps can be seen in the work carried out at the Medieval village site of 'VM4' located during the Boeotia Project surveys of the Valley of the Muses, located in central Greece (Bintliff 1996). Here archaeologists were faced with an overgrown hillside which surface pottery scatters had indicated was the location of a substantial Medieval settlement. Although some field terraces were clearly visible along with the foundations of a church, there were no obvious traces of the numerous house structures that would have been present at the site. This seemed to be the result of post abandonment robbing for the probable construction of sheepfolds and field boundaries, compounded by a dense mat of prickly oak obscuring large areas of the site environs.

Although the eyes could not penetrate the dense carpet of vegetation a surveying staff could, and in an attempt to locate any possible structural remains a total station was used to undertake a micro-topographic survey of the surface of the hillside. Once the data had been collected and collated, the GIS package Arc Info was used to construct a raster DEM (Figure 5.10 top left). As can be seen, whilst the general topographic trends are illustrated nicely, the DEM itself does not shed any further light upon the problem of the missing house platforms. To do this a slope layer was derived which is shown in Figure 5.10 (top right, bottom). A number of features are now evident and further processing through the application of a high-pass filter serves to further reveal clear groups of previously unsuspected house foundations (Gillings 2000).

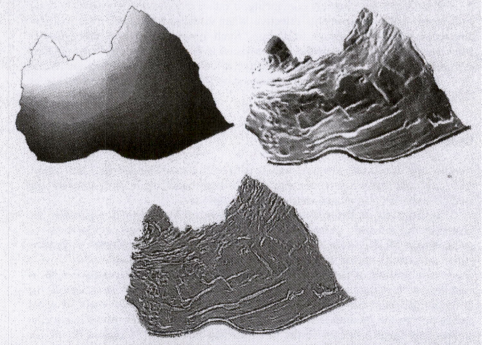

Figure 5.10 Top left shows the DEM derived from micro-topographic survey (scale ranges from white:high to black:low). Top right shows the slope map derived from the DEM and the bottom image shows the slope map having been high-pass filtered to further enhance the features.

Gillings (1995, 1997) has used a digital elevation model as part of The Upper Tisza Project, a large multi-disciplinary project investigating the Upper Tisza river in Hungary. During this project, understanding of the archaic landscape of the Upper Tisza river plain was hampered by extensive 19th century flood control works, and a lack of detailed quantitative records of flood events prior to these controls (Gillings 1997). To combat this, a detailed elevation model was constructed, and water runoff models were performed, allowing a simulated flood event to be reconstructed. Despite some reservations (due to the limitations of the elevation data) about the precise extent and location of the flood waters, this allowed some valuable conclusions about the character of the Neolithic landscape to be drawn:

"When looking at the shape of the flood event, the most striking feature of the simulated flood-zone was the way in which the observable landscape was partitioned by the intrusive flood-waters. Instead of giving rise to a simple wet block:dry block opposition, the seasonal flood dynamic appears to engender a much more complex restructuring of the observable socio-environmental space ..." (Gillings 1997).

5.6 VISUALISATION

Elevation models can productively be used to visually represent the variation of any quantity over space, and this has proved one of their most widespread applications within archaeological analysis. The visualisation of elevation data can itself be of considerable benefit to the understanding both of landform and of the quality of the elevation data itself. Visualisation can immediately reveal regular artefacts introduced by algorithms, survey instrument mis-calibrations or other forms of errors that would be far more difficult to perceive in other ways.

The usefulness of three-dimensional terrain representations is increased dramatically when different data themes are overlain on the elevation data. In this way it is possible to visually interpret the associations between, for example, fieldwalking data and terrain form. More recently, increases in the computational power available on the desktop have allowed *interactive* visualisation of terrain models. Here the calculation and rendering of the diagram is so rapid that it can be redisplayed fast enough to convince the eye that the terrain is moving, and input can be taken in the form of slider bars rather than typed numbers to complete the effect of interaction. Good examples of this are provided by the 'SG3D' tool, and 'NVIZ' (which run within the GRASS GIS on various platforms) and by Arcview's '3D analyst' extension. These programs take advantage of the advanced graphics acceleration of some workstations and PCs to display and animate wireframe, solid and even shaded renderings of elevation models in real time.

5.7 SUMMARY

In some sense this has been a rather odd chapter, essentially adding to the preceding sections on data structures (Chapter 2), data acquisition (Chapter 3) and manipulation (Chapter 4). At the same time, we have introduced a very specific case of *spatial interpolation*, which we will go on to discuss in much more general terms in Chapter 9. Our excuse for such a 'mixed bag' is the ubiquity and usefulness of terrain models, which argues strongly that they require a separate discussion.

We began by discussing the nature of elevation data in maps (touching upon elements of cartography) in order to illustrate some issues about the representation of terrain that are quite independent of spatial technology. We highlighted some of the problems and limitations inherent in any formal, discrete representation of a continuous variable such as topography, and drew particular attention to two methods: *contours* and *hachure plans*.

We then described the ways in which source elevation data, and then elevation models themselves can be stored within a GIS. In doing so we discussed the choice of representation—either altitude matrix or TIN—and highlighted some of the issues that arise because of this choice, particularly the importance of resolution.

The next section, on the creation of elevation models from contours, was deliberately a little more detailed. This was intended to draw the readers' attention to some of the limitations of elevation data that is derived in this way, and we hope it has encouraged a far more critical approach to elevation models than has been normal within archaeology. The potential differences in elevation models that are caused by the choice of algorithm should be borne in mind by any organisation which is creating or purchasing data that has been created in this way, and searching questions asked about the data. At its simplest, we would encourage all archaeologists to understand how their elevation data has been created, and then to ask of their data *is it fit for the purpose I intend it?*. This idea of 'fitness for purpose' is vital to the effective use of elevation models, because if the purpose is only to produce visual representations of terrain for quick appraisal then the absolute accuracy of the model, or the presence of systematic distortions, will be far less important than if the model is intended for analytical studies such as statistical, hydrological or visibility analysis.

Finally, we have looked at some of the immediate products of elevation models, drawing attention to some of the most obvious archaeological applications of these. In Chapters 9 and 10, we will look at two other methods that rely heavily on elevation models—cost surface analysis and visibility analysis—in considerably more detail.

CHAPTER SIX
Beginning to quantify spatial patterns

"Quantitative methods can properly be applied in a descriptive mode, as a way of untangling a complex of material without either implying causation or suggesting that the same outcome might recur." (Taylor and Johnston 1995: 61)

We have emphasised throughout this book that it is the spatial dimension of much archaeological data that makes it such a distinctive problem for data storage and manipulation. This spatial component also presents some distinctive problems if we wish to engage in quantitative data analysis, and this brings us to the area of *Spatial Analysis*. The use of spatial statistics in archaeology is one of the most difficult areas to introduce into a contemporary discussion of archaeological methods because it is so intimately associated with the 'new geography' and 'new archaeology' of the 1960s and 1970s. As a result, many archaeologists are either theoretically predisposed to reject the use of statistical approaches to space or are rapidly discouraged by the complexity of the statistics themselves.

This chapter attempts to address the second of these concerns by providing an overview of how some of the most widespread methods work without excessive formal notation, but the first of these issues deserves some comment. In Chapter 1 we explained that many contemporary theoretical approaches to archaeological interpretation emphasise the social construction of space or even the subjective experience of space in the past and present, and criticise the use of formal models as being reductionist or anti-historical. We are familiar with these issues but do not want to take a particular view in this chapter (which is *not* to say that we do not have a view). Instead, we argue that spatial statistics may still be useful to us regardless of how we theorise space.

In essence, we see formal spatial analysis not as a means of producing complete archaeological interpretations but as an extension of our observational equipment. Although the human mind is a fine interpretative tool, if it is presented with a series of random dots it does have a tendency to suggest patterns even if none exist. For example, when presented with a distribution map or a plot of stake holes from an excavation, the temptation to 'join the dots' is almost irresistible. When we do see patterns, more often than not these can resemble faces or other familiar objects. This is not a 'fault', of course, but a good way of recognising and remembering spatial patterns. We need only consider constellations of stars to see how convenient and necessary it is. However, it also means that we cannot rely entirely on our eyes and brains to give us an unbiased interpretation of things that exhibit spatial patterning, and it is here that quantification can be a great help to us. So we need not regard spatial statistics or formal models as *explanations* of the spatial arrangement of cultural remains in order to acknowledge that they have a part to play in our analyses. We might think of them as kinds of numerical and graphical 'prostheses' whose use enhances our ability to identify and describe spatial patterns in much the same way that we use binoculars and microscopes to overcome the physical limitations of our eyesight. None of this argues that graphical representation of our data is not a good thing. On the contrary, good graphs and charts should be a required element of most analysis and, in the case of Exploratory Data Analysis (see below), can actually form the basis of a reflexive and exploratory approach to multivariate analysis.

We have deliberately avoided the use of pages and pages of equations because the idea is to provide an overview of some popular methods and to describe how they work without becoming too bogged-down in the formal notation. This means that we do not provide 'complete' descriptions of most of the methods we discuss and where greater detail is needed then we will direct the reader to Shennan (1997) who provides a thorough coverage of the archaeological uses of classical statistics or to Hodder and Orton (1976), the archetypal work on spatial statistics in archaeology. Many other relevant publications can be found beyond the archaeological literature and, where appropriate, we direct readers to Bailey and Gatrell (1995), which gives an introduction to many more methods than we discuss here or to McGrew and Monroe (1993), who cover several of the classical statistical methods and provide some very useful worked examples.

6.1 WHAT IS SPATIAL ANALYSIS?

Spatial analysis comprises 'a set of techniques whose results are dependent on the locations of the objects of analysis' (Goodchild 1996:241). This encompasses a variety of techniques ranging from the statistically complex and computer intensive, down to the simple visual appraisal of a distribution map.

Through the course of Chapter 4 we discussed a wide-ranging set of tools and capabilities that enabled the GIS to routinely identify, extract and re-express the information held within the overall database. It can be argued that all fall within the rubric of spatial analysis, as defined above, but none of these techniques tell us whether any of the trends or patterns are *statistically significant* or not. A clear distinction can therefore be drawn between techniques that *identify* and *summarise* spatial patterns, and those which *investigate* these patterns. As a result of this, in a recent review of spatial analysis and GIS, spatial analysis was re-defined to draw an explicit distinction between techniques of spatial summarisation and those of spatial analysis proper (Bailey 1994:15-16). Inherent in the latter was the importance of statistical description and modelling. In this formulation the GIS capabilities we discussed in Chapter 4 such as query, reclassification and map algebra can be identified as spatial summarisation techniques—prerequisites of spatial analysis but not constitutive of it.

More recently the term 'Spatial Data Analysis' has been used to further distinguish between applications of conventional statistical tests to spatial data, *i.e.* where the objects being analysed could be moved in space without affecting the outcome of the test, and those statistical techniques which take into account the locational component of the datasets (Bailey and Gatrell 1995). This distinction has been elegantly illustrated in an archaeological context by Kvamme (1993). What should be clear from this discussion is that spatial analysis represents much more than query and map overlay, and that within it some very distinct sub-fields can be identified.

6.2 IDENTIFYING STRUCTURE WHEN WE ONLY HAVE POINTS

Points are frequently used in archaeology as shorthand for archaeological entities. They might be classes of site at a regional scale or artefact locations, a smaller scale of analysis. The structure of groups of point locations is the simplest geographic patterning we can consider and, although it is frequently visualised as a distribution map, this presents us with a problem because—as we have pointed out above—we cannot rely on visual perusal of point distributions alone if we wish to make well-founded assertions about the presence of and nature of patterns within point distributions.

Formal statistical methods allow us to address this problem by phrasing questions about point distributions in numerical form. In the simplest case, we may wish to answer the question *does this set of point locations exhibit any spatial patterning?* and if so *what form does the pattern take?*. More interestingly perhaps we might want to ask '*how confident can I be that this pattern is not random?*' Here we have to make it clear that no archaeological distribution is ever literally 'random' but that the social and taphonomic structuring that has produced the archaeological record may be such that the resulting points do not exhibit any identifiable spatial structure that would distinguish the distribution from a random one. Where this is the case, then it is not really appropriate to build interpretations on the apparent spatial patterns.

The most obvious form of such patterning is the case of point *clustering*, in which point locations tend to occur near to one another (Figure 6.1 right). However, point patterning is complicated by the fact that we might be equally interested in patterns of points that exhibit high levels of *order*, as in Figure 6.1 (left). Both patterns are equally different from the pattern that might be expected if an effectively random process (Figure 6.1 middle) was responsible for the point locations.

Although this may seem nothing more than a trivial property of Euclidian geometry, it is worth considering its archaeological importance, and the way in which it may guide us to other methods for spatial analysis. If the points represent settlements, then an ordered distribution might lead us, for example, to suspect that locations were chosen to maximise the extent to which the area of resources exploited by each site do not overlap. As such, a *central place* model may be appropriate to further explore the distribution, possibly through a spatial allocation approach (see Chapter 7). A clustered pattern, on the other hand, might suggest a different set of structuring principles, such as a tendency to locate the sites near to

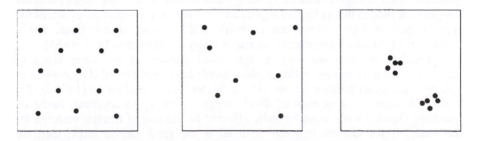

Figure 6.1 Types of point distributions: totally ordered (left), random (middle) and clustered (right).

raw materials or, if the points represent artefact locations, to deposit artefacts in special places.

Before we can begin to approach this, however, the first question we ask of point data should probably be *how confident am I that this pattern is non-random?* We have pointed out that our eyesight and brains cannot necessarily be trusted and so we may reasonably choose to use numerical methods to identify and measure the extent of randomness.

Quadrat methods

Quadrat analysis is one approach to the measurement of randomness within point distributions. This involves placing a regularly spaced grid over the distribution and comparing the number of points that occur in each grid cell. In a random population of points, the resulting frequency distribution of the number of points in each grid cell should theoretically be a Poisson distribution.

One property of the distribution of quadrat counts that can be exploited is the *Variance* to *Mean* ratio (*V/m*) of the distribution. The mean is simple to calculate (the sum of the number-of-points values divided by the number of quadrats) while the *variance* is a formal estimate of the dispersion of the distribution closely related to the *standard deviation* (Shennan 1997:41-42).

In a clustered distribution, this will tend to be large because the variance will be greater than the mean (due to higher than average numbers of points being found in some of the quadrats). For a dispersed distribution, the reverse will be true: the ratio will tend to be lower, because there will be little deviation from the mean. A value of close to one suggests a random arrangement of points. A test statistic can be calculated from:

$$\chi^2 = (V/m) \times (n - 1) \qquad\qquad\qquad (6.1)$$

(where *n* is the number of points) and can be compared with one for a random distribution of points with the same density using a χ^2 goodness-of-fit test.

This last point is important because merely describing the point pattern may not be enough. A variance to mean ratio of, say, 2.5 means very little on its own and we really need to qualify it by saying how unusual it is, expressed as *statistical significance*. For example, if the ratio is said to be 'significant at the 0.05 level' this means that it would only occur by chance once in every 20 times that we generated a set of dots at random. Although this is 'significant', we might put more weight on a pattern that is 'highly significant' at the 0.001 level; in other words the type of pattern that occurs once in every 1000 random distributions. For a discussion of statistical significance in archaeology see Shennan (1997:50-55).

Quadrat methods are easy to apply and convenient for some kinds of archaeological data, notably when artefact counts have been recorded by context or spit but their exact position has not. However, as Hodder and Orton (1976:36-37) point out, there are a number of disadvantages of using an approach based on quadrats. Quadrat studies can be badly affected by the size of quadrat selected for the study. If the quadrat unit size selected is too small or too large, then the variations between the numbers of points within each unit will be dramatically changed, and the result may suggest randomness or regularity when a clustered pattern is in fact present (Hodder and Orton 1976:37). Where clustering is present, a rule of thumb is that using quadrats that are approximately the same size as the clusters of points will most easily identify it, although this is little help when clusters are of widely varying sizes. Quadrat studies can also be adversely affected by the regular shape of the units—although in a GIS context these are almost always squares—and by the starting point of the quadrat grid.

To reduce these effects, it is a good idea to perform several quadrat analyses with differing sizes of quadrat, or by repeating several analyses with offset grid origins. It is also possible to use other properties of the point distribution to estimate an 'ideal' size for quadrat units or to use a 'moving window' (Bailey and Gatrell 1995:84) but none of these mitigation strategies overcome the inherent weaknesses of quadrat methods. A more detailed discussion of quadrat analysis, including a worked example, can be found in McGrew and Monroe (1993:222-227).

Nearest-neighbour distances

An alternative approach to the quantification of point-patterns can be based on *nearest-neighbour statistics*. These rely on the distance of each point to the nearest other point (usually denoted by r). In a region of area A, with n points, which has ρ points per unit area, then the mean nearest-neighbour distance of the points in a random point pattern will depend *only* on the density of points:

$$\bar{r} = \frac{1}{2\sqrt{\rho}} \tag{6.2}$$

The 'randomness' of the distribution can then be assessed by comparing the *expected* mean r (the value we would obtain if the distribution were really random), with the *observed* mean r (calculated from the actual distribution). This is usually expressed as R where:

$$R = \frac{\left(\bar{r}_{observed}\right)}{\left(\bar{r}_{expected}\right)} \tag{6.3}$$

In a random distribution of points R will obviously be close to 1. In a clustered distribution, however, it will be somewhere between 1 and zero and in a dispersed pattern it will tend towards its theoretical maximum of 2.1491 (this occurs in a 'perfect' dispersal pattern where the points form a triangular lattice). As with quadrat techniques, it is possible to construct a significance test for randomness using r. This uses the *standard error* (σ) of the expected $\bar{r}$ calculated as follows:

$$\sigma\left(\bar{r}_{expected}\right) = \frac{0.26136}{\sqrt{n\rho}} \tag{6.4}$$

and this can be used to generate a test statistic C as follows:

$$C = \left(\bar{r}_{observed} - \bar{r}_{expected}\right) / \sigma\left(\bar{r}_{expected}\right) \tag{6.5}$$

If the number of points is greater than about 100, this can be compared with the normal distribution so that, for example, $C = 1.96$ is significant at the 0.05 significance level, and $C = 2.58$ is significant at the 0.01 level.

Further enhancements of nearest-neighbour approaches are also possible by considering not only the distance to the nearest neighbour, but also by the distance to the second-nearest, third-nearest and so on.

Although not entirely without problems, particularly with respect to the imposition of boundaries on a point pattern, nearest-neighbour analyses generally produce a more robust assessment of patterns in a point pattern than quadrat analysis. Nearest-neighbour approaches are discussed in more detail in Hodder and Orton (1976:38-52) where some worked examples are also provided of their application to archaeological data and in much more detail by Bailey and Gatrell (1995:88-94). The approach has been used with some success within archaeology by, for example, Clark, Effland, and Johnstone (1977).

Dispersion of clusters around the mean centre

If clustering can be shown in a point distribution, then some measure of the dispersion of individual clusters can be obtained. This is analogous to the calculation of the mean and standard deviation in a non-spatial data set and conveys broadly the same information.

The mean centre of a distribution of points can be found by taking the mean of the *x* co-ordinates and the mean of the *y* co-ordinates. The resulting location is the mean centre. A circle, centred on this point just large enough to include two thirds of the points is referred to as one standard distance (see Figure 6.2). As an abstract measure, this is of little value in itself, but it can be of interest in the comparison of several clusters of sites or artefacts.

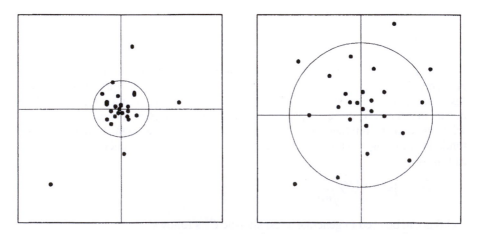

Figure 6.2 Low dispersion around the mean centre (left) compared with a wide dispersion of points (right). In each case the line represents the mean centre of the distribution and the circle encloses the same proportion of the points.

6.3 SPATIAL STRUCTURE AMONG POINTS THAT HAVE VALUES

Frequently, we have measurements or observations that go with our points. In the case of sites, we may have estimates of size, status or date while if we are dealing with artefact distributions we may have an almost infinite range of observations or measurements. We may, of course, still be interested in treating the locations of these artefacts in the same way as we have up to now, but we may also be interested in the distribution of their attributes. If so, then one of the first considerations is likely to be whether the values of the attributes actually exhibit any spatial patterning. Again, there are two ways in which a variable may show a pattern across space, and these are closely related to the three cases of clustering, ordering and randomness that we have discussed in relation to point distributions.

It is easiest to think about spatial pattern in points with values if we consider what we can deduce from an individual value about the values near to it. If knowing the value at one location helps us to guess the values of nearby points, then our points exhibit a property called *spatial autocorrelation*. This can take two forms: if neighbouring locations tend to have similar values to each other, then the spatial variable is said to exhibit *positive spatial autocorrelation*. On the other hand, if they tend to be different, (in other words the presence of a high value at one place makes it more likely that nearby values will be low) then the variable is said to exhibit *negative spatial autocorrelation*. The difference between this is similar to the example of clustered or ordered distributions of simple points. Spatial autocorrelation can actually be present in any kind of data, from point data to categorical data (such as geological maps) through to continuous spatial variables such as elevation. Indeed the existence of spatial autocorrelation in so many spatial variables is the basis of *regionalised variable theory*, *geostatistics* and *Kriging*, which we discuss further in Chapter 9.

The amount of spatial autocorrelation in a variable is therefore a description of the extent to which there is structure within the variable, and identifying spatial autocorrelation is often the first step in an analysis of some spatial variable. If archaeological data can be shown to be spatially autocorrelated, then there is something that needs to be explained within the data.

There are many ways of measuring and testing for spatial autocorrelation. One of these which has been widely discussed in relation to archaeological applications is Moran's *I* statistic. This is the only measure we will discuss here, but the reader may wish to consult Cliff and Ord (1973) for the most widely cited treatment of spatial autocorrelation including alternative statistics such as Geary's *c*, or Hodder and Orton (1976:181-184) for a discussion of both of these measures in an archaeological context. Moran's *I* can be defined as:

$$I = \frac{n \sum_{i=1}^{n} \sum_{j=1}^{n} w_{ij}(x_i - \bar{x})(x_j - \bar{x})}{\sum_{i=1}^{n} \sum_{j=1}^{n} w_{ij} \sum_{i=1}^{n} (x_i - \bar{x})^2} \tag{6.6}$$

where the values are denoted by *x*, and $\bar{x}$ is the mean value of the variable. Admittedly this looks like a rather complex equation, but in fact the idea is relatively straightforward: each observation x_i is compared with each other observation x_j. The spatial relationship between them is expressed with a weighting term w_{ij}, which is arranged so that it is large for pairs of points that are close to each other and decreases with distance between points. In practice this term is usually the inverse of the distance between the locations being compared ($1/d_{ij}$) or sometimes the inverse of the *square* of distance ($1/d_{ij}^2$), which reduces still further the impact of more distant points. Regardless of the particular weighting term chosen, Moran's *I* theoretically varies from -1 (for 'perfect' negative spatial autocorrelation) to +1 (for 'perfect' positive autocorrelation). It is therefore a useful measure of the extent to which a given distribution of sites or artefacts is exhibiting spatial pattern.

Just as with our quantifications of point distributions, the value of Moran's *I* alone is not easy to interpret. Once again, it is more useful to us if we can also say that our result for Moran's *I* is *significant* at a particular level. A significance test for Moran's *I* can actually be constructed in two ways: either by assuming that the x_i observations are drawn from a normally distributed population (called the 'normality assumption') or by considering the calculated value of *I* relative to the set of all possible permutations of *I* with *n* observations (this is called the 'randomisation assumption'). In practice, the results of a significance test based on these assumptions rarely produce different results, but the randomisation assumption is sometimes argued to be more useful in archaeological situations, because the characteristics of the population cannot usually be ascertained. Kvamme (1990d) provides an archaeological discussion of the use of Moran's *I* that also includes a worked example of significance tests under both of these assumptions. Some care has to be exercised as to whether we are interested in identifying only positive spatial autocorrelation, in which case we must construct a significance test as a *one-tailed* test, or whether we are interested more generally in whether the data shows any spatial autocorrelation, either negative or positive, in which case a *two-tailed* test is appropriate.

If positive autocorrelation can be shown to exist within our archaeological observations, then we are justified in beginning to interpret this variation. In some situations we may also then choose to explore their distribution further by using trend surface analysis (see the boxed discussion of the Maya Collapse data set) or we might wish to exploit the existence of spatial structure through one of the methods of interpolation, both of which are described in Chapter 9.

6.4 SPATIAL STRUCTURE IN AREA AND CONTINUOUS DATA

Although point locations are the most often used form of data within archaeology, it is not unusual to have archaeological observations that refer instead to areas. We may, for example, have a database of the archaeological characteristics of political units such as parishes, counties or countries. This kind of data is particularly frequent where our source is a Sites and Monuments Record or other inventory of archaeological sites. At other times archaeological data may have been recorded so that the observations have been grouped together into contexts, squares or other recording units, and the original locations of the observations are not known. Maps and visual representations of such data are made considerably easier using GIS than they would have been in the past, but still suffer from the problems of pattern-identification that we have discussed above. Fortunately, however, there are numerical methods for identifying spatial autocorrelation in this kind of data also.

SPATIAL AUTOCORRELATION AND THE STRANGE CASE OF THE MAYA COLLAPSE: A CAUTIONARY TALE

A good example of the use of spatial autocorrelation measures can be found in the literature on Classic Lowland Maya settlement sites (see Bove 1981, Whiteley and Clark 1985, Kvamme 1990d, Neiman 1997).

The end of monument construction activity at these sites can be dated—albeit imperfectly—by the date of the latest monument known to have been constructed at them. All the sites seem to date to a relatively short period beween about 750AD and 900AD, a period sometimes rather dramatically referred to as the 'Classic Maya collapse'. This series of dates has been used by a number of archaeologists (e.g. Bove 1981 and more recently Neiman 1997) to generate 'trend surfaces' in order to test ideas about how these changes in Maya settlement were caused. Several possibilities have been proposed, including the idea that pressure from the western frontier of the Maya region initiated the apparent decline. Such analyses obviously all assume that the Maya data preserve some spatial structure with respect to these dates—in other words that the dates associated with the sites exhibit *spatial autocorrelation*.

However, the existence of spatial autocorrelation in the Classic Maya data was called into question when Whiteley and Clark (1985) tested the distribution of dated monuments using a Moran's *I* statistic. Rather surprisingly, their test seemed to show that there was *no* spatial autocorrelation present in the data—a conclusion that would have rendered the trend surface analyses, among other things, meaningless. The problem was subsequently resolved by Kvamme (1990d), who reviewed Whiteley and Clark's methodology and concluded that they had calculated a version of Moran's *I* that was designed for area data.

This is important because, to calculate their statistic, they had been compelled to impose an arbitrary grid over the sites which aggregated the available data from 47 sites into only 17 grid squares. This constituted a massive reduction in the sample size and reduced the test's ability to detect spatial autocorrelation if it were present. When Kvamme applied a revised, and much more appropriate, *point* data version of Moran's *I*, significant spatial autocorrelation could indeed be detected and Bove's trend surface analysis was vindicated.

This example therefore provides a good example of the usefulness of calculating a numeric index for spatial autocorrelation, but also sends us two warnings. Firstly, it is important to use the appropriate statistic for the kind of data we have: Whiteley and Clark's analysis was flawed not because they miscalculated their statistic, but because they selected a procedure that had little inferential power for their particular data. Secondly, we should avoid over-interpreting a negative statistical result: we need to remember that a negative result (such as Whiteley and Clark's) really says '*no spatial autocorrelation was detected*', rather than '*there is no autocorrelation present*'.

Figure 6.3 Spatial autocorrelation in county and raster data. Extreme cases of positive (left) and negative (right) spatial autocorrelation in county (in this case squares) data.

One approach is to use a formulation of Moran's *I* that is specifically intended for this kind of data. In situations where 'county' data is being used then the weights W_{ij} in Equation 6.6 can be set to 1 for the case where x_i and x_j share a border, and 0 for the case where they do not. Sometimes the proportion of x_i's border that it shares with x_j is used to modify this crude weighting so that a value of Moran's *I* can be calculated, and a significance test established in a similar way as for point data.

Spatial variables that exhibit continuous variation are usually dealt with in GIS by representing them as raster data (although see Chapter 5 for discussion of *e.g.* Triangulated Irregular Networks) and in this case there are also ways of calculating Moran's *I*. Here, the weights can be arranged to account for the distance between the grid cells in a number of different ways. The grid cells might be treated as areas, just as above, in which case the weight will be 1 for grid cells that are adjacent in one of the cardinal (N,S,E,W) directions and zero for all others. This can be modified to the extent that grid cells on the diagonals obtain a weighting above zero as well as the grid cells in the cardinal directions, or even to include 'knights move' distances between grid cells. Alternatively, the grid cells can be treated in a similar way to point data, by using the distance between the centres of each pair of grid cells to derive a weight based on the inverse of the distance.

6.5 STRUCTURE IN LINES AND NETWORKS

There are actually many occasions in archaeology when we may want to represent our information as connections between point entities, and it is these connections that we are actually interested in rather than the absolute spatial position of the points. Lithic artefact refitting and sites linked by transport routes are obvious examples of this kind of information and there may be very good reasons for thinking that representing the world as a network rather than Euclidian geometry might be theoretically desirable. Networks may actually be a better representation of the 'model' of the world that people retain within their minds. We know this intuitively every time we explore a new town or landscape because we construct our mental representation of the area by remembering places (nodes) and the routes between them (arcs) rather than by constructing a map of their relative positions in Euclidian space. Confirmation of this comes when we are suddenly surprised that two places which our mental map will only allow us to reach from a common—possibly quite distant—third location are in fact adjacent and easy to move between. The success of the famous London Transport Underground map is based entirely on the representation of a set of points—in this case tube stations—and the connections between them as a network that bears no necessary relationship to their physical configuration.

Remembering our discussion of the spatial database in Chapter 1, what we are discussing here is analysis of the *topological* component of the data rather than the spatial component. This can be obvious and tangible, as in the case of road networks, rail networks and rivers, or it can be more theoretical and abstract, as in the case of sites connected by epigraphic evidence or trade connections. Any observation that can link together two places can be used to define a network and

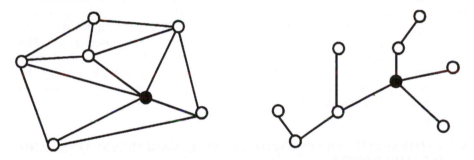

Figure 6.4 A highly connected network (left) for which b = 1.857 and all the Konig numbers are 2; and a tree network (right) for which b = 0.889. In this case the black node (Konig number = 3) is clearly more 'central' than the grey one (Konig number = 5).

in fact one of the largest application areas for the analysis of topology and networks is *social network* analysis, which involves no consideration of a spatial component at all (see *e.g.* Wasserman and Faust 1994).

Despite some obvious advantages, and an historic interest by new geographers in network analysis (Hagget and Chorley 1969), archaeologists have shown a very limited interest in networks as a medium for formal analysis of spatial relationships. Apart from occasional applications of location-allocation modelling (Bell and Church 1985, Mackie 1998) and some limited applications of network models to trade patterns (Allen 1990, Zubrow 1990a), the use of networks has tended to be restricted to the analysis of domestic spaces, as exemplified by Hillier and Hanson's 'Social Logic of Space' approach (Hillier and Hanson 1984). The increasingly widespread application of vector-GIS which have the inherent capability to undertake these kinds of analyses, suggest that a re-appraisal of network models is now quite likely.

Network analysis is really just a geographer's term for a subset of a branch of mathematics called *graph theory* (see *e.g.* Wilson 1985 for an introduction). In network analyses, systems of lines are represented in terms of *arcs,* which end at *nodes.* The arcs may only be connected at the nodes.

From the discussion of topology in Chapter 2, it should be clear that vector-GIS are, in many ways, ideally suited to the analysis of networks as these are the elements we discussed in relation to storage of vector data. In GIS terms, we are discussing here the analysis of the topological component of the database rather than the spatial component. If represented in this way, geographical patterning in phenomena such as road and river systems can be explored in detail, generating descriptive statistics or developing algorithmic approaches to analysis.

For example, it is possible to generate various statistical indices to represent the *connectivity* of a network, the *centrality* of nodes within the network and the *spread* of a network. High connectivity within a network implies that many of the nodes are linked together, and a simple index of the connectivity b can be obtained by dividing the number of arcs by the number of nodes. This varies between 0 for networks consisting of only nodes, to more than 1 for highly connected networks. Figure 6.4 shows how this index reflects the structure of two 'connected' networks (without isolated nodes). Counting the minimum number of arcs, by the shortest path, which must be traversed in order to travel from a node to the node farthest from it, provides the *Konig number* of a node. This is a useful indicator of the centrality of any given node within the network. In Figure 6.4 (right) the black node has a Konig number of 3 while the grey one has a Konig number of 5. The *diameter* of a network can be defined as the minimum number of arcs, which must be traversed to move between the farthest two nodes and, as such, will always be the same as the highest Konig number of the network. This can then be used to make indices of the spread of a network.

6.6 COMPARING POINTS WITH SPATIAL VARIABLES: ONE- AND TWO-SAMPLE TESTS

So far we have only considered structure within a particular data theme, either points, areas or networks. Commonly we may be interested in whether one spatial object exhibits any patterning with respect to another. Examples of this are commonplace within the analysis of archaeological distributions: we may wish to know whether a particular group of archaeological sites is preferentially distributed on particular soils, or with respect to some social convention about space or, within a single site, we may wish to compare the distribution of lithic deposits with proximity to hearths.

Here we may choose to undertake tests of association, of which there are a great many. Fortunately, GIS are ideal for producing the values necessary to undertake these kinds of tests and, as we will see, they actually present some inferential as well as practical advantages in identifying significant patterns of association.

Given a set of archaeological locations, the desire to know whether or not they are randomly distributed with respect to some spatial phenomenon traditionally requires the construction of a *two-sample significance test* of association. For this, the characteristics of the archaeological locations are tabulated against the spatial variable; a random sample of locations from the same region is generated, and the same characteristic recorded for these locations as for

EXAMPLE: TWO-SAMPLE TEST OF ASSOCIATION

A useful test for two-sample data of ordinal level or higher is Mann & Whitney's U test (see *e.g.* Shennan 1997:65). This has several useful characteristics, particularly that it does not assume normally distributed data and it allows comparison of two samples of different sizes. For samples x and y of size N_1 and N_2 respectively, values of U_1 and U_2 can be calculated as follows:

$$U_1 = N_1 N_2 + \frac{N_1(N_1+1)}{2} - \Sigma R_x$$

$$U_2 = N_1 N_2 + \frac{N_2(N_2+1)}{2} - \Sigma R_y \qquad (6.7)$$

where R represents the rank order of an observation within *both* samples: in other words ranked from 1 to (N_1+N_2). Where either of the samples is greater than about 20, the U statistic can be considered to be normally distributed and a z score can be calculated from the following:

$$z = \frac{U_1 - (N_1 N_2 / 2)}{\sqrt{[N_1 N_2 (N_1 + N_2 + 1)/2}} \qquad (6.8)$$

In a two-tailed test, and for a significance level $\alpha = 0.05$, z must therefore exceed 1.96 to reject H_0, and for the difference in the samples to be considered significant.

Table 6.1 shows four columns, the first two of which are directly derived from a GIS study of long barrows, and represent the scores of a series of 27 monuments (barrows) on some spatial variable (in this case an index of the visibility of the barrows). The second column represents a series of 30 random points, treated exactly as the long barrows, and scored on the same index. Rx and Ry then represent the rank order of the observations in the entire series of 27+30 observations.

The U statistics calculated from the equations 1.7 are $U_1 = 322.00$, and $U_2 = 488.00$. This provides a z-score of 1.33, which is not significant at the 0.05 level. This suggests that it is not possible to state that the distribution of the barrows x is non-random with respect to the index of visibility - of course there *may* be a pattern but this test does not provide the evidence for it.

the archaeological locations.

Two opposing hypotheses are constructed to test whether archaeological sites are non-randomly located with respect to the distribution of the characteristic or not. The null hypothesis (the hypothesis that they *are* random, designated H_0) is then tested. To do this, the two samples are compared with each other in order to ascertain how likely it is that they were drawn from the same statistical population. If, at the chosen confidence interval, they could have been drawn from the same population then H_0 holds, and the archaeological locations cannot be said to be non-randomly distributed. If they seem to be drawn from different populations then

H_0 is rejected and it can be said that the sites are non-randomly distributed with respect to the characteristic (Shennan 1997:48-70).

GIS can rapidly produce all the information necessary to conduct a two-sample test by generating random point locations from the same geographical region as the archaeological locations and then rapidly assigning the spatial variables to both the archaeological locations and random sites. A test appropriate to the nature of the data can then be undertaken. There are many suitable tests including the *Mann and Whitney test* (see boxed example) and two procedures that are frequently used in archaeology, the χ^2 test (Shennan 1997:104-126) for

Table 6.1 Example of a Mann-Whitney U-test for 27 long barrows (x) compared to 30 random points (y) by generating the rank order within the entire series of each observation (Rx and Ry).

Mounds (x)	Random (y)	Rx	Ry
7465	7401	57	56
7065	5146	55	52
6879	4903	54	51
5360	3869	53	46
4678	3823	50	45
4276	3774	49	44
4026	2926	48	36
3962	2882	47	35
3251	2829	43	34
3235	2614	42	32
3199	2567	41	31
3153	2548	40	30
3111	2367	39	28
3020	2309	38	27
2968	2240	37	26
2652	2176	33	25
2501	2135	29	24
2112	1914	23	22
1800	1823	20	21
1384	1779	17	19
1194	1456	14	18
1152	1373	13	16
972	1339	9	15
968	1041	8	12
707	1025	4	11
311	1011	2	10
216	924	1	7
	796		6
	743		5
	586		3

categorical data (*e.g.* to test whether a particular distribution is random with respect to soil class, or geology) and the *Kolmogorov-Smirnov test* (Shennan 1997:57), which is appropriate when the spatial variable is ordinal or higher—for example elevation, slope or rainfall.

As Kvamme (1990c) has pointed out, however, two-sample approaches are hampered by the fact that the characteristics of the population from which the samples are drawn are not directly observed. This may not be a problem with large samples because the larger the samples, the more accurate the estimate of the population characteristics will be and the more likely the test will be to identify an association if one exists. However, because GIS can produce the values of a spatial variable for an entire geographic region—either by treating the grid cells as a population, or by calculating the areas of vector polygons—we might choose to regard those as the parameters of the population itself. In that case, we can directly compare the characteristics of the archaeological locations with the characteristics of the population and undertake a potentially more powerful *one-sample significance test*. There are one-sample versions of most of the tests of association that we may wish to conduct, including the *t*-test, the χ^2 and Kolmogorov-Smirnov tests.

6.7 RELATIONSHIPS BETWEEN DIFFERENT KINDS OF SPATIAL OBSERVATIONS

There are also many archaeological situations in which we may want to consider the spatial distribution of some measurement or observation that we have obtained at particular sites, for example we may wish to consider the distribution of the average size of flakes at different manufacturing sites in a region in relation to distance from the source of the raw material. In this case we are effectively looking for correlations between two measured variables and there are two kinds of associations that we may expect to find.

Where a high value of one variable generally implies a high value in another we can speak of *positive correlation* between the variables. This is the case where, for example, two types of artefacts are commonly associated on archaeological sites: finding one will make it more likely that the other will also be present.

We may be equally interested, however, in cases of *negative correlation* between our spatial variables, the case where a high value in one variable implies a low value in another, as might be the case where artefacts have mutually exclusive distributions.

One of the most popular measures is *Pearson's r* (more correctly called *Pearson's product-moment correlation coefficient*), which is a measure of covariation between two variables measured at interval or ratio scale data. This is a relatively simple calculation to make from a table of paired values, using the formula:

$$r = \frac{\sum XY - \left(\sum X\right)\left(\sum Y\right)/N}{\sqrt{\sum X^2 - \left(\sum X\right)^2/N}\sqrt{\sum Y^2 - \left(\sum Y\right)^2/N}} \qquad (6.7)$$

EXAMPLE: ONE-SAMPLE TEST OF ASSOCIATION

Table 6.2 shows the output from a GIS, structured for the construction of a one-sample *Kolmogorov-Smirnov* test. In this case the spatial variable in which we are interested varies from 0 to 16 and is mapped for the entire region. The population values are shown in columns 2 and 3, with the cumulative frequency shown in column 4. Column 5 shows the number of mounds that occur in each of these classes (also obtained from the GIS), and columns 6 and 7 show this converted into a cumulative frequency for comparison with the population values.

In a Kolmogorov-Smirnov test, the maximum difference (D_{max}) between the two cumulative frequencies is used to determine whether or not the sample deviates significantly from the population.

In a one-sample case, at a significance level of 0.05, the critical value that D_{max} must exceed is calculated from:

$$d \approx 1.36 \Big/ \sqrt{n} \tag{6.8}$$

which in this case is 0.26. From the values in Table 6.2, it can be seen that the maximum difference between the cumulative frequencies occurs at value 3 on visibility index variable, and that this does not exceed the critical value of 0.26 for the chosen significance level.

This test also, therefore, fails to allow rejection of the null hypothesis that the barrows are randomly distributed with respect to this index of visibility, although—as with the two-sample example—we must be careful not to interpret this as proof that no association exists. Indeed, our sample size of 27 is really too low for the test to be reliable (it is intended for samples of 40 and upwards) and so we should not be surprised that it is not successful. In fact, we can see from inspecting the graph that the two cumulative curves are not identical so there is a pattern in this data, even if it is not so marked as to be statistically significant.

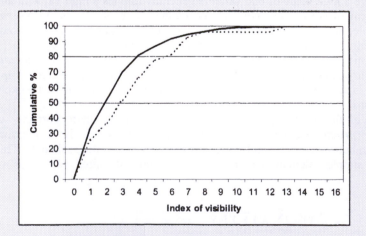

Figure 6.5 The cumulated percentages from Table 6.2 shown as a graph. The solid line represents the population, while the dashed line is the cases.

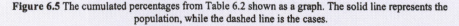

Table 6.2 One-sample Kolmogorov-Smirnov test: the test compares the distribution of the population (derived from GIS maps) and the cases (long barrows) with respect to the index of visibility.

| | Background population | | | Archaeological cases | | | |
	Area km^2	Area%	Cum%	Cases	Cases %	Cum%	D
0	0.00	0.00	0.00	0	0.00	0.00	0.00
1	133.08	33.27	33.27	7	25.93	25.93	0.07
2	75.97	18.99	52.26	3	11.11	37.04	0.15
3	70.40	17.60	69.86	4	14.81	51.85	*0.18*
4	45.65	11.41	81.27	4	14.81	66.67	0.15
5	22.17	5.54	86.82	3	11.11	77.78	0.09
6	19.03	4.76	91.57	1	3.70	81.48	0.10
7	10.89	2.72	94.30	3	11.11	92.59	0.02
8	8.31	2.08	96.37	1	3.70	96.30	0.00
9	6.47	1.62	97.99	0	0.00	96.30	0.02
10	3.72	0.93	98.92	0	0.00	96.30	0.03
11	1.86	0.47	99.38	0	0.00	96.30	0.03
12	1.02	0.25	99.64	0	0.00	96.30	0.03
13	0.77	0.19	99.83	1	3.70	100.00	0.00
14	0.42	0.11	99.94	0	0.00	100.00	0.00
15	0.17	0.05	99.99	0	0.00	100.00	0.00
16	0.04	0.01	100	0	0.00	100.00	0.00

and a test statistic t can be calculated using:

$$t = \frac{r\sqrt{n-2}}{\sqrt{1-r^2}} \qquad (6.9)$$

This can be compared against standard tables of t to form a test of significance. Although the use of Pearson's r is ubiquitous in geography, and is not uncommon in archaeology, some caution should be exercised in using it as it requires several assumptions to be met, notably that the variables have a linear association, that they are measured at interval or ratio scale and that the variables are normally distributed. See *e.g.* McGrew and Monroe (1993:252-254) for further details and a worked example.

An alternative test for data measured at the ordinal scale is *Spearman's rank correlation coefficient* (McGrew and Monroe 1993:254). This makes less assumptions than a test using Pearson's r, assuming only that we have a random sample of paired variables, which have a monotonic association (increasing or decreasing—but not necessarily linear). It does not assume that the samples are drawn from normally distributed populations—an assumption that is rarely supportable in spatial analysis.

It can be calculated from:

$$r_s = 1 - \frac{6\sum d^2}{N^3 - N}$$ (6.10)

where N represents the number of paired values, d is the difference in ranks of variables X and Y for each paired data value and d are the corresponding differences in ranks. Again, a fuller explanation and worked example can be found in McGrew and Monroe (1993:256-258).

6.8 EXPLORATORY DATA ANALYSIS

"Pictures that emphasize what we already know—'security blankets' to reassure us—are frequently not worth the space they take. Pictures that have to be gone over with a reading glass to see the main point are wasteful of time and inadequate of effect. **The greatest value of a picture** is when it forces us to notice **what we never expected to see.**" (Tukey 1977: vi, emphases in original)

Above, we discussed how the introduction of formal spatial statistics into archaeology was in a large part due to perceived subjectivities involved in 'eye-balling' distribution maps, a practice that has historically dominated a great deal of archaeological investigation, and we cautioned above against placing too much trust in our eyes. However, there is one important branch of statistics that actually places a premium upon visual examinations of data. These techniques are grouped under the general banner of *Exploratory Data Analysis* (EDA) although they have to date only found sporadic application with archaeology.

EDA is an approach to statistics that was pioneered in the 1970s by the statistician John W. Tukey. In identifying the dominance and acknowledging the utility of existing confirmatory data analysis, Tukey highlighted the fact that such approaches tend to follow a basic pattern. Firstly, a hypothesis is generated (to give a common archaeological example *site location is positively correlated with areas of good soil*), then a model of the relationship is fitted to the data, statistical summaries are obtained (*e.g.* means, standard deviations, variances) and finally these are tested against the probability that such values could have occurred by chance (Hartwig and Dearing 1979:10). Tukey saw this process as inherently restrictive, insofar as only two alternatives are ever considered (either the sites are correlated or they are not). It also places far too much trust in statistical summaries that can have hidden assumptions about the data they summarise, or obscure and ignore vital information. What was needed—to precede and then run alongside traditional confirmatory data analysis—was an exploratory phase of analysis. A strong principal underlying Tukey's EDA was that analysis should always begin with the *data* itself, not a summary of it.

As a result EDA encourages researchers to effectively re-frame their questions. For example, rather than asking *is site location positively correlated with areas of good soil?* they should instead ask *what can the data in front of me tell me about the relationship between site locations and soil types in the study area?*.

Rather than submit a body of data to a single confirmatory statistical test, researchers are instead encouraged to search the data, using a variety of alternate techniques to assess it. The underlying assumption is that the more you know about the data the more effective you will be in developing, testing and refining theories based upon it.

In addition, one of the most powerful characteristics of EDA is the emphasis that is placed upon data visualisation in this exploratory process, where a variety of visual displays of the data should be used in advance of, and alongside, traditional statistical summaries.

In seeking to extract the maximum information from a given data set Hartwig and Dearing have identified two fundamental principles of the EDA approach. These they have termed *scepticism* and *openness*. Researchers should be *sceptical* of statistical measures that summarise data (*e.g.* the mean) since they may conceal or misrepresent what may be the most informative aspects of the data under study. In addition they should also be *open* to the possibility of unanticipated patterns appearing in the data (Hartwig and Dearing 1979:9). What is clear is that to advocates of EDA the hypothetico-deductive statistical paradigm, or 'confirmatory mode of analysis' that dominates the statistical approaches we have discussed so far in this chapter is neither sufficiently *open* nor *sceptical*.

A very wide range of EDA techniques exist and interested readers are referred to Tukey (1977), Hartwig and Dearing (1979) and StatSoft (2001) for excellent, accessible and thorough discussions of EDA methods. In the remainder of this section we will look at one technique that has been successfully utilised in archaeological-GIS based studies and emphasises the strong visual character of EDA techniques. This makes extensive use of graphical representations (icon plots) that rely upon bringing human brains' full visual processing capabilities to bear upon the task of pattern recognition and exploration.

Icon plots are a powerful form of multivariate exploratory data analysis (StatSoft 2001). The underlying rationale of icon plots is to represent an individual unit of observation (or case) as a particular type of graphical object. In archaeology this graphical object has most commonly been the star. In the case of star-plots each case is represented as a point with a number of radiating arms corresponding to the number of variables that need to be incorporated into the analysis. This can encode presence/absence or the length of each arm can be used to represent continuous data, where the length is proportional to the value of each variable for each specific case—usually scaled between 0 and 1. The exploratory value of such plots has been neatly summarised by StatSoft, Inc:

"The assignment is such that the overall appearance of the objects changes as a function of the configuration of values. Thus, the objects are given visual 'identities' that are unique for configurations of values and that can be identified by the observer. Examining such icons may help to discover specific clusters of both simple relations and interactions between variables." *(StatSoft 2001)*

An example is given in Figure 6.6, taken from a study of the relationship between prehistoric site locations and notions of risk and economic potential in a dynamic floodplain environment (Gillings 1997). Here there are three arms reflecting the following environmental variables: distance to a drainage basin interface (BASIN or 'B'); distance to the edge of the flood zone (FLOOD or 'F'); distance to the streams and rivers that would have channelled flood water (HYDRO or 'H'). Visual study of the resultant star-field for a series of 143 sites enabled a series of clear classes to be identified suggesting a variety of responses to the flood-plain environment (Figure 6.7 and Table 6.3).

Table 6.3 Classes of response identified through the visual analysis of the resultant star-plot field
(adapted from Gillings 1997).

Class	Flood risk	Economic potential
1	HIGH	HIGH
2	HIGH	LOW
3	LOW	LOW
4	LOW	HIGH

Star-plots were also used successfully in a pioneering evaluation study undertaken in the late 1980s on the US military base of Fort Hood. Here the star-plot technique was used to investigate early historic period (1900-1920) structures found in the study area (Williams *et al.* 1990:252). In practice stars were used to summarise the presence/absence of variables relating to certain architectural features and although a wide range of star-types were produced, clear trends could be identified within this, such as the presence of core-attributes such as chimneys.

A variation on the star-plot approach that indicates the emphasis placed upon visual data representations by the EDA approach is the use of Chernoff faces, shown in Figure 6.8. This is very similar to the star-plot but instead of using star shape and size to encode information, makes use of the human capability to recognise and respond to very subtle changes in the shape and expression of human faces. Instead of star-arms variable values control the size, shape and overall expression of a stylised human face (Chernoff 1973).

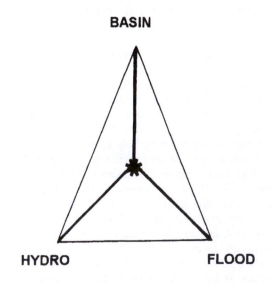

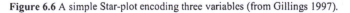

Figure 6.6 A simple Star-plot encoding three variables (from Gillings 1997).

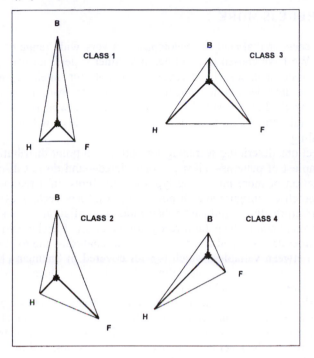

Figure 6.7 Star-plot archetypes for the flood-plain sites (from Gillings 1997).

As yet the full potential of EDA has not been developed within archaeological-GIS with very few published case studies upon which to draw. Despite this we believe that EDA offers a powerful conceptual approach to statistical analysis and a wide range of practical techniques that enable researchers to better explore and understand their data and we strongly recommend that readers follow up the references given in this chapter.

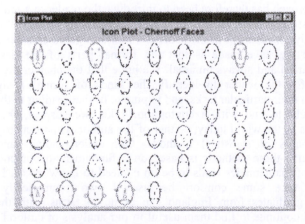

Figure 6.8 Chernoff faces (StatSoft, Inc 2001).

6.9 AND THERE IS MORE ...

As we have seen, spatial analysis encompasses a very wide range of methods and approaches. We have chosen to describe, in a rather superficial way, only a small number of basic methods. In defence, it is not our intention to provide a comprehensive introduction to statistical methods or even to spatial analysis (however defined), but we would hope that the reader who is not familiar with these areas might now be more inclined to learn, or at least will now know where to begin reading.

The sections describing searching for patterns in point distributions consider only the simplest of patterns—clustered or ordered—and do not discuss methods for identification of more interesting types of structures. In a real archaeological situation geometric configurations of points (circles, lines, rectangles) are likely to have more interpretative significance than random configurations and this can be dealt with through automatic pattern-recognition (see *e.g.* Fletcher and Lock 1980). We have deliberately stopped short of regression analysis—the exploration of the relationship between variables, which is well covered in Shennan (1997) both in bivariate and multivariate forms, although regression does crop up in our discussion of predictive modelling in Chapter 8.

Similarly, we have made no mention of multivariate statistics that might be of considerable relevance to spatial analysis (cluster analysis, principal components analysis, correspondence analysis and others). For these, the reader should first ensure familiarity at the level that Shennan (1997) provides, and then could not be better advised than to refer to Baxter (1994).

6.10 SPATIAL ANALYSIS?

This chapter has been difficult to write because we have been, to be colloquial, caught between a rock and a hard place. A book such as this on spatial technology would clearly not be complete without a chapter discussing formal spatial analysis (although a frightening number have been written) but, at the same time, to do justice to the subject would require an entire book on its own.

Our approach has therefore been to introduce the reader to a fairly wide range of basic statistical procedures that most archaeologists might expect to understand without massive investment in mathematics, matrix algebra or other formal skills. The sacrifice we have made is that the formal parts of the descriptions are sometimes partial and always minimal. We hope that the chapter provides enough information for the reader to identify the kinds of formal analysis that are appropriate to their particular tasks, and to pursue further reading in those areas.

We ought to end with a confession. Like, we suspect, the majority of archaeologists in the world today, neither of the authors is particularly well versed in formal maths or statistical methods. Our understanding—and hence our presentation of—these methods is based on a fundamentally practical approach, and our use of mathematics tends to be on a 'need to know' basis. From this, the reader may take some comfort because if two 'numerically challenged' archaeology graduates from England can come to terms with these kinds of methods, and continue to find them useful in our analysis of cultural remains then there must be hope for all of us.

CHAPTER SEVEN

Sites, territories and distance

"The catchment of the site was linked to its economic contents, which included animal bones and carbonised cereal grains. Site catchment analysis there provided, in an appropriately circular fashion, the explanation for site location." (Gamble 2001: 145)

Distance is the most fundamental property of spatial data. It is the fact that proximity, or distance from one another, may have a direct influence upon the attributes or relationships between things that makes explicitly geographic observations different from other types of data. Proximity and distance are also at the core of many important archaeological questions. The task that archaeology sets itself as a discipline is to explain the material remains of the past, and this clearly includes a desire to explain how things came to be where they are, and incidentally, are not in any of the other places they might have been.

As with most of the interpretative archaeology discussed in this book, attempts to explain the spatial distribution of cultural remains considerably predate the availability of GIS. In fact, archaeology and anthropology have such an established tradition of theories and methods for spatial analysis that some have tried to justify the definition of a specific sub-discipline of 'spatial archaeology' (Clarke 1977a). In practice, much of the theory and method for explaining spatial organisation has come to us via geography, which in turn 'borrowed' many of its models from a range of disciplines including physics, economics, biology, ecology and pure geometry. Particularly notable in the development of spatial archaeology is the 'New Geography' (*e.g.* Haggett 1965) which clearly influenced the adoption of similar approaches to archaeological materials (*e.g.* Hodder and Orton 1976).

In seeking explanations for the spatial configuration of cultural remains, archaeologists, like geographers, tended to concentrate on the distribution of archaeological sites—usually settlement sites—and to this end turned to a number of theoretical approaches. Prominent among these have been gravity models (Hodder and Orton 1976:187-195); von Thunen's (1966) economic model of settlement structure; Christaller's 'Central Place' theories of settlement hierarchy (Christaller 1935, 1966) and papers in Grant (1986a), and ecologically-based resource concentration models (Butzer 1982).

Since the 1970s the need for an explicit 'spatial archaeology' declined as these techniques slipped into the methodological mainstream of the discipline. There was also a gradual realisation that for all of their methodological rigour, in many cases the formal analysis of space offered little more than a sophisticated *description* of patterns rather than explanation of them (Hodder 1992).

More recently, as a result of changes in the theoretical climate broadly termed post-processual, the centrality of the body and complexity and historical specificity of the concept of space itself have been highlighted. This has led to the development of a range of more qualitative new approaches to the study of human spatiality (*e.g.* Tilley 1994). It is interesting to note that this has been undertaken without the need to resurrect an explicitly defined 'spatial archaeology'.

As we will show, the advent of spatial technologies such as GIS has prompted a resurgence of interest in the quantitative techniques characteristic of the spatial archaeology of the 1970s, allowing archaeologists to introduce far greater sophistication into the formal analysis of space. Spatial technology provides some extremely useful methodological building-blocks through which quantitative approaches to the spatial arrangement of cultural remains can be approached. Spatial technologies, including GIS, may yet prove to be the catalyst for the re-definition and emergence of spatial archaeology as a discrete topic of study. The challenge will be to ensure that both the formal techniques introduced by the New Archaeology and the more qualitative approaches characteristic of more recent developments in archaeological theory fall within its rubric.

7.1 BUFFERS, CORRIDORS AND PROXIMITY SURFACES

The most basic, and at the same time one of the most useful abilities of GIS is the generation of distance products, either in the form of proximity buffers, or continuous proximity surfaces. The simplest case of this type of product can be considered to be *proximity surfaces*, which are—at least in theory—expressions of a function in which the magnitude at any point of the map is the measured proximity to a particular geographic entity or entities. It is worth noting that whilst in the majority of current applications this proximity corresponds to a quantified distance, there is nothing to stop us also integrating more cultural factors. For

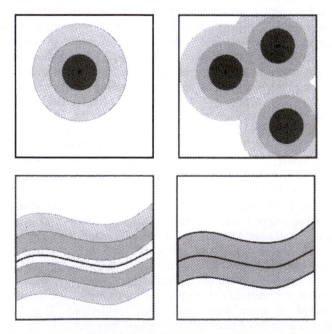

Figure 7.1 Buffers and corridors. Distance buffers from a single point (top left), distance buffers from several points (top right), distance buffers from a line (bottom left) and a corridor from a line (bottom right).

example, regardless of the measured distance on the ground, a settlement may always be deemed 'close' if it is in-view whereas a tomb may likewise be felt as 'distant'.

Within vector-GIS the most straightforward distance products may be termed distance *buffers* and *corridors*. These are categorical products in which the classes represent a range of proximities to some source entity or entities. The classes may define regular intervals of distance from source—for example 0-10km, 11-20km and so on, although buffer operators can just as easily be configured to provide uneven, entirely different ranges more suitable for particular analyses (*e.g.* 0-500m, 500m-5km, 5km-100km). The term *buffers* is usually reserved for a map with several distance bands generated from either points or lines, while the term *corridor* is sometimes used to specify the case where a single distance buffer is generated from a line (see Figure 7.1).

The process of generating a buffer or corridor involves the selection of the source features (points, lines or polygons) and specification of the desired buffer width, as is the case with systems such as Arcview. Distance buffers and corridors have been widely used in archaeological analysis to construct hypothesis tests (see Chapter 6) regarding the distribution of archaeological sites or find spots. The key advantage that is offered by GIS in this respect is that the system can automatically determine the area covered by each of the bands. This is useful as, statistically, the areas of the bands can be regarded as the parameters of the statistical population from which the locations of archaeological sites can be regarded as a sample.

Within raster systems, the definition of buffers and corridors is a two-stage process. In the first instance a distance layer is calculated with respect to the source cells we would like to buffer. Here each cell in the resultant layer is given a value which corresponds to its distance from the source cells. The second stage is to then reclassify the distance layer so that all cells whose value falls beneath the predetermined buffer threshold are given a value of '1' (in buffer) and the remainder '0' (out of buffer). The same technique can be used to establish multiple bands if required. In saying this, some predominantly raster systems such as Idrisi now offer the same push-button functionality as vector systems.

7.2 VORONOI TESSELLATION AND DELAUNAY TRIANGULATION

The generation of distance buffers or corridors can be regarded as a form of *spatial allocation*, and it would therefore be remiss not to expand that discussion to include the other main method of spatial allocation that has been used within archaeology - *tessellation*. Tessellation splits up the available space into tiles based on the geometric properties of a point distribution.

The most commonly used tessellation within both geography and archaeology has been the *Voronoi* tessellation (also called Thiessen Polygons), and this is closely related to another graph that can be generated from a set of points: the *Delaunay Triangulation*. This approach was introduced to archaeology by the 'New Archaeology', notably David Clarke (Clarke 1968, 1972) and has frequently been used in archaeology to define notional site 'territories'. Well-known examples of its application include the definition of territories for Neolithic Long Barrows (Renfrew 1973:545) or for Iron Age Hillforts (Cunliffe 1971, Grant 1986b) and the analysis of Romano-British Settlement (Hodder 1972).

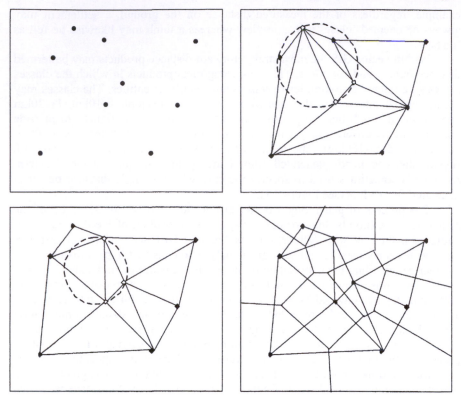

Figure 7.2 Delaunay triangulation. The set of points shown top left can be used to create many different triangulations, including the example shown (top right). The Delaunay triangulation (bottom left) is unique, because circles whose circumference passes through the points of the triangles contain no other points. It can also be used to create the Voronoi tessellation (bottom right).

Whereas tessellation refers to the creation of a set of geometric areas that have the source points located within them, triangulation refers to the creation of a graph in which the points are used to form the corners of a set of triangles. For any given set of points, a large number of possible triangulations are possible depending on which points are joined to which others. The *Delaunay* triangulation of a set of points is defined as that in which the resulting triangles are closest to equilateral, and in which the circles whose circumferences pass through the points of the triangles contain no other points. This is shown graphically in Figure 7.2.

The Voronoi tessellation is related to the Delaunay triangulation because it can be created by (1) projecting lines from the midpoints of the triangle edges and at right angles to them and then (2) using the intersection of these lines to define the polygons. By definition, Voronoi polygons contain only one of the source points but they also have the very useful property that they enclose all of the space that is closer to that point than any others.

The close association of this method of spatial allocation with the functionalist interpretative stance of early New Archaeology has led to the technique becoming far less used in archaeological analysis today although the recent popularity of GIS has perpetuated its use somewhat (see *e.g.* Savage 1990, Ruggles and Church 1996).

Aside from the theoretical reasons, there are some more methodological reasons why Voronoi tessellation is not currently a widely applied method of spatial allocation. These centre on the fact that the tessellation *is* a solely geometric allocation method and, as such, it assumes that the social and geographic features of an area have no impact on the allocation of space. In response to this, where spatial allocation is regarded as a useful analytical method, instead of simple tessellations, contemporary archaeologists have more frequently turned to variants of what is termed *cost surface* modelling.

7.3 COST AND TIME SURFACES

Distance products, however useful, are based solely on the Cartesian spatial properties of the data. In settlement studies, this need not be a problem if we are content to make simple analogies between cultural and physical systems, and if we expect our results to represent some 'ideal' solution for use as a comparator to the actual pattern of settlements observed. However, it becomes problematic if we wish to develop methodologies that take account of obstructions, barriers and differences in the qualities of space that may have influenced transportation costs, movement or even perception of the landscape.

Two locations that are the same linear distance from a source location may not be equally easy to reach: for one it may be necessary to walk uphill, while there may be only flat terrain to cross to reach the other. Equally, some areas of the landscape may be deemed socially 'dangerous' or 'taboo' and may exert powerful negative influences upon movement—for example liminal zones such as burial grounds. Consequently, it will be necessary either to travel slower or to expend more energy (or mental resolve) in reaching one point than the other and we can argue that it is, in many circumstances, more useful to develop models of the *time taken* or the *cost* incurred in reaching any given target location from a source location rather than the distance.

Many GIS offer the facility to generate *cost surfaces*. These can be regarded as modifications to the continuous proximity product that take account not only of proximity but also of the character of the terrain over which that proximity is measured. Of course, like distance, these are still simply mathematical models whose archaeological meaning is not fixed: it is up to us to use these as building-blocks to create methodologies that have some archaeological meaning.

There are a variety of slightly different algorithms implemented in different GIS but generally they can be of two types: *isotropic* algorithms, which take account of the cost of movement across the surface but take no account of the direction of movement or *anisotropic* algorithms in which direction is considered to affect the cost of movement across terrain[5]. It should be noted that cost surface

---

[5] Thanks and apologies to Tyler Bell, who DW once mistakenly tried to convince (in a bar in Barcelona) that the correct definition was entirely the opposite of this. The term derives from 'iso' (equal) 'tropic' (directional) and 'anisotropic' therefore refers to algorithms that are *not* equal with respect to direction. The terms 'isotropic' and 'anisotropic' are often used to describe algorithms or approaches that may or may not account for direction, for example anisotropic semivariograms and Kriging (see Chapter 9).

operators are almost always performed on raster-based data rather than vector, although there is no technical reason why cost surface routines could not be written for use with vector-based terrain models such as triangulated irregular networks (see Chapter 5). As a result it is commonly the case that predominantly raster-based systems (such as Idrisi and GRASS) have comprehensive and powerful cost surface tools built in as standard.

Isotropic cost surface analysis

For the simplest case of isotropic cost surface calculation, the algorithm requires two inputs: a file containing the location of the features from which cost distance will be calculated, often called the *seed locations* and a file which contains the cost of travel across each landscape unit usually called a *friction surface*. Most algorithms expect this to contain values for the proportional cost of crossing each landscape unit relative to a notional base cost of 1. Thus a location coded 0.5 would incur half the effort to traverse as the base cost, and a location coded 3 would require three times the effort. With these two layers as input, the algorithm begins at each of the seed locations and successively generates the costs of reaching neighbouring raster cells. In this way, it cumulatively models the *cost* of reaching any point in the landscape from one of the seed points.

It is important to note that cost surface algorithms do not inherently model anything 'real' such as movement or dispersion. On the contrary, *cost* is rather an abstract concept, and the results of cost surface analysis can be used to represent a number of different outputs. Most often, cost surface analysis has been used to model either the *time* taken to move between a source and other parts of the landscape or the *energy* expended in moving from the source and the other locations.

Anisotropic cost surface analysis

Isotropic analysis may be quite acceptable in situations where return-to-base cost is being analysed, or in cases where the direction of movement through some friction does not affect the cost incurred (or the time taken) in crossing it. However, there are cases in which the direction of movement across a friction makes a difference to the cost incurred.

One such case is slope—a cost is incurred in moving uphill (in other words the friction should be positive and greater than one), while a *benefit* might be accrued if we are moving downhill (in other words the cost ought to be fractional). In this case, it may be necessary to use a more sophisticated cost model to generate the cost surface. One approach is to create inputs for both the magnitude of the friction (as before) but also the direction in which the friction has its greatest effect. Some equation is then used to calculate the *effective friction* caused by traversing a location from (1) the direction of movement (relative to the friction's direction) and (2) the magnitude of the friction.

There are several methods of implementing an anisotropic function, but to show how it works we will discuss the algorithm implemented in the Idrisi GIS package (Eastman 1993), from which much of the following description is drawn.

It uses the following equation to calculate the actual effective cost of traversing any given cell:

$$Cost = Cost_{base} \times F_{max}{}^{\cos^{k}(\alpha)} \tag{7.1}$$

where $Cost_{base}$ is the base cost, F_{max} is the maximum magnitude of the friction, α is the angle between the direction of travel and the direction of the friction and k is a constant used to effect the costs between $\alpha = 0°$ and $\alpha = 180°$.

To see how this works, we can assume that the *base cost* is 5, the friction due to maximum slope is 2, and that $k = 1$ (the effect of k will be explained later). When the direction of movement is in effect directly upslope (*i.e.* $\alpha = 0$):

$$Cost = 5 \times 2^{\cos(0)} \quad \text{or} \quad Cost = 5 \times 2^{1} = 10 \tag{7.2}$$

which is twice the base cost of traversing the cell. However, when the value of α exceeds 0—in other words when travel is not the same as the aspect of the terrain—then different results for cost are obtained. When $\alpha = 90°$ (when movement is directly *across* the slope) then the equation produces an effective cost as follows:

$$Cost = 5 \times 2^{\cos(90°)} \quad \text{or} \quad Cost = 5 \times 2^{1} = 5 \tag{7.3}$$

which is the same as the base cost. This too is what we would expect, because movement across a slope neither gains or loses any height. Where movement is down the slope (*i.e.* where $\alpha = 180°$), the equation works out as follows:

$$Cost = 5 \times 2^{\cos(180°)} \quad \text{or} \quad Cost = 5 \times 2^{-1} = 2.5 \tag{7.4}$$

which is *half* the base cost of the cell, modelling the fact that moving downhill involves less than the base cost.

Between the two extremes, the equation produces intermediate results, allowing for movement in any direction across a cell. At $\alpha = 45°$, for example (movement which is slightly uphill, and slightly across the hill) the equation produces the following:

$$Cost = 5 \times 2^{\cos(45°)} \quad \text{or (approximately)} \quad Cost = 5 \times 2^{0.7} = 8 \tag{7.5}$$

or an effective friction of 0.8 which implies (as expected) some additional cost over the base cost, but not as much as for movement directly uphill.

The constant k is used to 'tune' the way that slope modifies the friction values. Figure 7.3 shows the effect of increasing the value of k on the effective friction. The top line is the case discussed above, where $k = 1$. Here the value of the effective friction at $\alpha = 45°$ is about 0.8 (as shown above) but increasing the value of k as shown to 2, 10 or 100 progressively increases the effective friction between the two extremes. This allows the relationship between slope, direction of movement and cost (or time) to be modelled in quite a sophisticated manner.

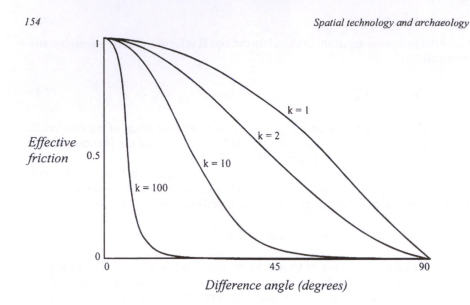

Figure 7.3 Relationship between difference angle and effective friction, as modified by the *k* parameter (after Eastman 1993).

The friction surface

The creation of the friction surface entirely determines the nature of the analysis, and as such the derivation of the friction values deserves very careful consideration. Van Leusen (1999) has published a useful review of algorithms that can be used to derive friction values, and we draw extensively on this in the following discussion.

Where the output is intended as a model of energy, then the friction surface must represent the energy expended in traversing a fixed distance. If we assume, for simplicity, that the energy expended is solely a product of the slope of the terrain then a relationship must then be postulated between slope and energy. We might note, in passing, that this itself could be regarded as a worrying simplification, and will return to this issue.

Even if we merely consider the relationship between slope and energy, we can immediately note some complications. Although this relationship can be justifiably simplified, in reality it is asymmetrical (uphill slopes and downhill slopes of the same value do not have the same effect) and non-linear. As a result a variety of different equations have been used in archaeological situations.

A simple and robust approach is to start from 'backpackers equations' such as that of Ericson and Goldstein (1980) who estimate effort as follows (where H is the horizontal distance travelled, D^{up} is positive vertical height change and D^{down} negative vertical height change):

$$Effort = H + 3.168D^{up} + 1.2D^{down} \tag{7.6}$$

Some care should be taken to ensure that a particular 'backpacker's equation' is appropriate, as one derived from Himalayan climbing is unlikely to provide an appropriate estimate of energy expenditure in a rainforest or on a pavement.

Other authors have used data derived from physiological studies. Llobera (1999) used physiological experimental results (Kamon 1970, Minetti 1995), and extrapolated beyond these using data from mountaincraft manuals to arrive at a relationship between slope and energy applicable for most situations (Figure 7.4). Bell and Lock (2000:88) estimated the energy needed to move a body of given mass a given height although, as Bell and Lock themselves point out, this neglects the fact that moving downhill can be as tiring as moving uphill because of the constant braking that is needed to prevent 'uncontrolled descent' (falling).

Where cost surface analysis is intended as a time model, then the friction component needs to represent the time taken to traverse each cell, expressed as a proportion of the base cost for traversal of a cell. If we again accept the simplification above, then a relationship between slope and time is needed to create a friction surface. This equation can again be derived from 'backpackers equations', of which several have been published such as that of Gorenflo and Gale (1990), which estimates the speed by the exponential function (where v is velocity and s the slope in percent):

$$v = 6e^{-3.5|s+0.05|} \tag{7.7}$$

The vast majority of archaeological applications have so far accepted the simplification that energy expended or time taken to move around in a landscape is a function of slope. We observed above that this is a worrying oversimplification, and now draw attention to differences in energy expended moving over different terrain types, for example soft sand as opposed to paved road or through different vegetation covers, such as dense rainforest, swamp or open savannah. Marble (1996) has suggested the use of an equation that includes a term for ease of movement over terrain (where m is the energy expended in watts, w is the weight of the individual, l the weight of any load carried, v is the speed of travel, g the gradient in percent and n is the terrain factor):

$$m = 1.5w + 2(w+l) + \left(\frac{l}{w}\right)^2 + n(w+l)(1.5v^2 + 0.35vg) \tag{7.8}$$

It is important to realise that there can be more to 'cost' surfaces than merely slope and physical exertion because patterns of movement within a landscape are influenced by both physical and symbolic resources. Avoidance of cemeteries or desire to pass through villages, for example, may have had more influence on movement patterns than the nature of the topography. It is also worth noting, however, that 'respect for the ancestors' or 'sense of attachment to place' equations are somewhat rarer than those for backpacking. This should not be taken to suggest that such considerations should not be incorporated into our cost surfaces—this is one of the real challenges for GIS to rise to.

Whichever assertion is chosen about the relationship between friction and slope, it is essential that it is clearly specified with the results. Cost surfaces published without stating the derivation of the friction surface must be treated with

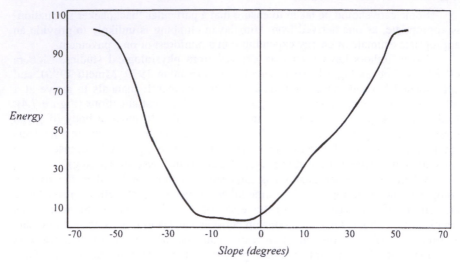

Figure 7.4 Relationship between energy and slope deduced from a combination of physiological studies and educated guesswork (redrawn from Llobera 1999 and reproduced with permission). Note that the relationship is neither linear nor symmetrical and that least energy is expended traversing a mild downhill slope.

extreme scepticism, as the form of the result is entirely dependent on the choice of friction values. Moreover, it is advisable to 'calibrate' the values if possible as was done, for example, by Gaffney and Stancic (1991).

Barriers, transportation routes and terrain

Apart from the relationship between terrain type and energy expended, movement can be affected adversely by barriers, such as rivers, fences, or cultural 'no-go' zones, or positively by the availability of transport routes. Some algorithms allow barriers to be explicitly coded into the friction surface with a value that will prevent movement across it (this has to be a value, such as a negative number, that could not occur as a friction value) while others do not. Where no explicit code is provided for barriers, the algorithm can usually be 'fooled' into behaving in the same way by inserting an extremely high value for friction. So long as the value is so high that it is cheaper for the algorithm to travel around the obstacle, then the result gained will be similar.

Similarly transport routes can be coded by including extremely low values of friction. For example, we may include a road or pathway along which we predict that movement could be twice as rapid as over unmade ground. In this case, we would create a map theme in which the pathway was coded as 0.5 and the rest of the theme as 1 and we would multiply the friction theme by this before using it.

Problems occur when we want to model a feature as both a barrier *and* a transportation opportunity. The most obvious example of this are rivers, which present a substantial barrier to movement across them, but can provide rapid transport up and down if a canoe is available. Even in this case, however, there are ways of modelling the system using ingenious coding of a friction surface. One possibility is to represent the river as a central corridor of very low friction, with a suitable value to represent the rapid low-energy transportation up and down a river, but to surround that with a thin buffer of high cost to represent half the cost involved in, for example, acquiring transportation or crossing the river.

Least-cost pathways

Using the result of a cost surface algorithm, it is possible to ascertain the 'least-cost' pathway between any given location in the landscape and the source of that cost surface routine. In most situations, this is exactly analogous to the hydrological runoff models discussed briefly in Chapter 5—the algorithm regards the cost surface as an elevation model whose lowest point (zero) is the target location—in effect the drainage point of the study area. It then simulates the effect of a drop of water being placed at each of the remaining cells in the layer and plots their passage to the target.

Before turning to the archaeological application of these, it is important to note—as has Douglas (1994)—that due to the quantization effects of modelling movement locally within a grid-based model, such algorithms are far from optimal in their calculation of these pathways. Cost surface algorithms themselves are

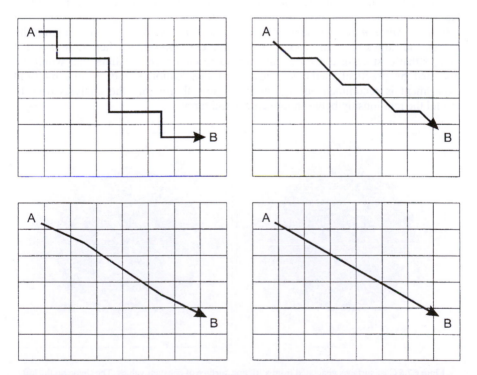

Figure 7.5 'Least-cost paths' generated in a uniform environment by a 4-connected (top left) or an 8-connected (top right) algorithm. If 'knights move' searching is allowed, then the distortion (bottom left) is more acceptable, although none generate the real optimal solution of a straight line (bottom right).

prone to significant errors caused by the fact that they are constrained to 'move' from any given cell in the raster grid in either 4 or 8 directions. This problem is even more significant in the generation of least-cost pathways because unless the source and target are aligned in one of these directions, then the distance travelled between source and target will be severely impacted by this.

This point is illustrated by Figure 7.5, in which the simplest case of a uniform plane is used to illustrate how different algorithms can react. As Douglas has subsequently asked of an incredibly simple example such as this: "if an algorithm fails to locate anything other than a straight line, then how do you expect it to work in a non-uniform plane?". (Douglas 1999)

Least-cost pathways have been employed in a number of archaeological analyses, for a detailed recent example readers are referred to Bell and Lock (2000) who used such calculations to analyse the prehistoric route called the 'Ridgeway' that cuts across southern England.

More methodological considerations

Just as with visibility analysis and any other analytical method that relies heavily on a digital elevation model, it is important to consider the effect on the analysis that errors and inaccuracy of the DEM may have. Considerations will include both the absolute accuracy and precision of the elevation model, where inaccuracies will lead to erroneous generation of slope estimates and therefore friction values, but also the *resolution* of the model. Where the resolution is not sufficient to permit the slopes and ridges of a landscape to be adequately represented, then there is a real danger that cost surface results will bear no resemblance to values that would be obtained by practical methods. Because of this, calibration of results by field experiment is recommended regardless of how thoroughly the analysis tries to model the relationship between movement and terrain.

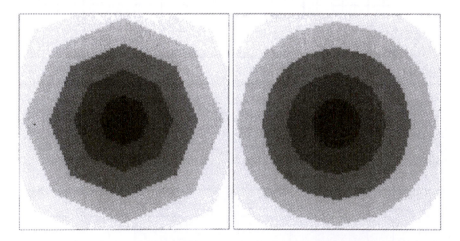

Figure 7.6 Cost surfaces generated from a friction surface of constant values. The shape on the left results from cumulated error in distance estimates between diagonals and cardinal directions, the diagram on the right illustrates the improvement obtained by using the 'knights move' algorithm.

It is also worth noting that the algorithms implemented in many commercial GIS provide far from 'ideal' estimates of cost surfaces. To illustrate this, we should consider the 'perfect case' of a uniform friction surface. The cost of reaching any location from a seed location will therefore be dependent solely on the Cartesian distance and a cost surface algorithm should produce a series of perfect circles— exactly the same result as a distance buffer.

In practice, most algorithms work iteratively by cumulating the cost in local neighbourhoods. Although they may correct for travel in a diagonal direction, there will always be a cumulated error in the estimated distance travelled, and hence cost incurred, at locations that are not reached directly along the diagonals, or in the cardinal directions. The effects of this can be seen in Figure 7.6 (left).

Some systems compensate for this by allowing the algorithm to estimate the 'knights move' distance, and this makes a considerable improvement (as can be seen in Figure 7.6 (right) but even with this correction, the algorithm cannot generate perfect circles.

Although cost surface analysis is not a perfect tool for modelling human movement in a complex environment, it can and has been used with some success within archaeology. However, more complete resolution of the problems that we have identified may require a more sophisticated simulation-based methodology— see various authors in Hodder (1978), also Lake (2000, In Press). In this case, it is necessary to begin programming new modules for GIS rather than relying on the application of standard toolkit functions and this lies outside the scope of this introductory text.

7.4 SITE CATCHMENT ANALYSIS AND GIS

Site catchment analysis (Vita-Finzi and Higgs 1970)– which we will refer to as SCA for brevity—is a technique for analysing the locations of archaeological sites with respect to the economic resources which are available to them. Site catchment analysis derives from optimal foraging theory. The basic tenet of the method is that the farther from the base site resources are, the greater the economic cost of exploiting them. Eventually there is a point at which the cost of exploitation outstrips the return and an economic boundary can be defined at this point to define the *exploitation territory* of a site (see *e.g.* papers in Findlow and Ericson 1980). Such studies have been criticised for the highly functionalist stance that they adopt. While this is not the appropriate place to discuss the merits or otherwise of 'optimisation theories', readers should be aware that models which assume that early human communities functioned as simple optimisers are open to considerable theoretical criticism—*e.g.* Gaffney *et al.* (1996b).

The capacity of GIS for extracting a variety of information about the environment and performing both geometric and statistical operations on it have led to its use for the application of SCA in archaeology (Gaffney and Stancic 1991, Hunt 1992). As we will see below, the capabilities of GIS can be used to add considerably to the methodological sophistication of SCA.

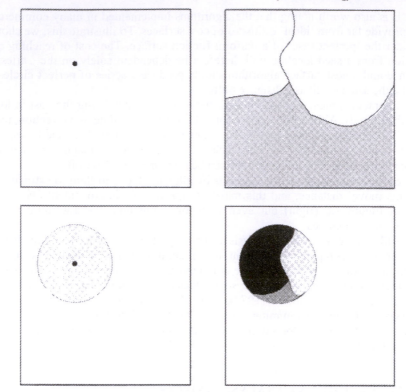

Figure 7.7 A simple Site Catchment Analysis implemented using GIS distance and overlay operations.
The site (top left) and the resource map in which we are interested (top right) form the inputs. The sites
are buffered (bottom left) and then crosstabulated with the resources map (bottom right).

At its simplest, the exploitation territory or *catchment* of a settlement or base
camp can be approximated as a circular area centred on the site in question.
Estimates of where the boundaries of these territories should be placed have been
derived from ethnography and differ depending on the economic base of the
communities. For example, the economic catchment of a sedentary agricultural site
may be placed at a 1 hour walk while for a herding/hunting community it may be 2
hours (Bintliff 1977:112).

Once the site territory has been defined, the proportions of given resources
within this area can easily be obtained using simple GIS map-algebra operations
(see Figure 7.7). These can be analysed and tabulated for comparison with the
resources within the catchments of other sites. In this way, it is claimed, the
differences between the sites may become apparent. Take for example a regional
distribution of contemporaneous sites. Some correspond to settlements used for
cereal growing while others are bases for hunting activities. As a result, the
resources which are available to the agricultural sites (as represented by the
characteristics of their catchments) might be expected to exhibit higher proportions
of lighter, fertile soils.

Although distance is an easy measure to calculate and use for the estimation

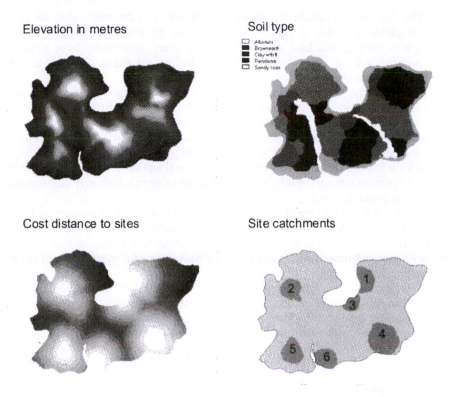

Figure 7.8 The digital elevation model (top left) from which the friction surface is derived (not shown), and the map of soil types (top right). Bottom left shows the result of the cost surface calculation, which can be reclassified into a catchment map (bottom right) for the sites.

of site territories or exploitation areas, it takes no account of the nature of the terrain surrounding a site. Consequently, no allowance is made for the fact that some parts of circular 'territories' are actually far more inaccessible than others, perhaps because of the slope of the terrain, or the presence of roads or paths, or the presence of cultural or physical obstacles such as 'taboo' areas or rivers. We will see below how one way to model the effects of these factors is to generate the territories based on the *cost distance* rather than the geometric distance from the sites.

The best way to illustrate how GIS can offer a more sophisticated form of site catchment analysis is through a hypothetical example. The example presented here is a fictional island, on which there are a series of sites which have been interpreted as Mesolithic seasonal camps. A digital elevation model is available for the island, and a map of soil types classified into five classes: alluvial soils; brownearths; clay soils; rendsinas; and sand soils (which occur predominantly around the coast). This will be used to suggest the different economic resources which are available to each site. In a real example we would expect to give more consideration to the representation of the economic resources, but for clarity we will assume that this rather unusual collection of soil types are a good indicator of the economic resources.

The first stage in calculating the site catchments is to determine how the cost function will be derived. In our hypothetical case, we will assume that we are dealing with the seasonal base sites of hunter-gatherers, who would have foraged for resources over a given period of time, before returning to the site for distribution and consumption.

Consequently, we are interested in the 'return to base' cost of each trip, so we are justified in using an isotropic function: if there is a difference between difficulty of the outward and return journeys, we can assume that these will generally cancel each other out over the entire foraging trip. For simplicity, we will make another assumption that slope will have the greatest effect on accessibility, and derive a friction surface from this (again, in a real example we would probably require a far more complex model of the cost of movement which may incorporate pathways or obstacles *etc.*).

The friction data theme is obtained by first generating the slope surface from the elevation map (this operation is discussed in Chapter 5), and then performing the selected transformation of slope surface according to a proposed relationship between slope and time taken. In our case, we have defined $F = 1/\log(\text{slope}+1)$ for illustration purposes only.

The cost distance layer is then re-classified according to the equation to identify zones which are within two hours walk of the sites. This process is summarised in Figure 7.8. The relative proportions of the various soil types occurring in those zones can then be tabulated and compared for the six source sites (Table 7.1).

7.5 CONCLUSION

In this chapter we have explored one of the most fundamental properties of spatial data—distance—and looked at the ways in which distance products can be determined and put to use in our GIS-based analyses. In doing so we have discussed a number of analytical approaches such as site catchment analysis and allocation-based studies that have witnessed something of a methodological renaissance as a direct result of the adoption of GIS. In doing so there has been a very deliberate sub-text to this chapter. While we hope to have shown the methodological 'value' which has been added to such approaches by the GIS it is

Table 7.1 Results of the crosstabulation, showing the sites (rows) and the area of each soil type (in this case in km^2) in that site's catchment.

	Alluv	Brown	Clay	Rends	Sand	Total
1	0.00	0.15	0.86	0.12	0.55	*1.68*
2	0.32	0.84	0.55	0.02	0.01	*1.74*
3	0.08	0.31	0.32	0.00	0.13	*0.84*
4	0.95	1.37	0.00	0.98	0.30	*3.60*
5	0.00	0.34	0.74	0.90	0.01	*1.99*
6	0.00	0.18	0.00	0.10	1.41	*1.69*
Total	*3.38*	*19.81*	*6.24*	*19.81 13*	*.97*	*11.54*

important to realise that they are not free of more fundamental theoretical criticism and that in adopting them we must accept that thirty or so years of intense development in archaeological thought has taken place since their first introduction. As a result we have attempted to stress the importance of more cultural variables and would urge GIS practitioners to explore the challenges that integrating such considerations into our analytical environments raise.

CHAPTER EIGHT

Location models and prediction

"Predictive models are tools for projecting known patterns or relationships into unknown times or places" (Warren and Asch 2000:6)

One of the most frequent applications of GIS has been in the construction of what have come to be known as *predictive models* of archaeological resources. Prediction is a central component of spatial analysis, and the goal of predictive modelling is to generate a spatial model that has predictive implications for future observations. In archaeological contexts, this means that the aim is to construct a hypothesis about the location of archaeological remains that can be used to predict the locations of sites that have not yet been observed. Kvamme puts it as follows:

"... a predictive archaeological locational model may simply be regarded as an assignment procedure, or rule, that correctly indicates an archaeological event outcome at a land parcel location with greater probability than that attributable to chance." (Kvamme 1990a:261)

As with most of the analytical techniques we have discussed, the approach of predictive modelling predates the widespread use of GIS by archaeologists. Much of the work on the development of archaeological predictive models was undertaken in the United States throughout the 1970s and 80s when various government agencies became interested in the prediction of the locations of archaeological sites within fairly large regions from information based on surveys of far smaller areas. A series of seminal discussions relating to the development of predictive modelling can be found in the volume 'Quantifying the Present and Predicting the Past' published in 1988 by the US Department of the Interior Bureau of Land Management (Judge and Sebastian 1988). This is a volume that any GIS practitioner interested in exploring the potentials of predictive modelling would be well advised to acquaint himself or herself with.

The main impetus for the development of predictive methods has therefore been practical and economic: it is held by proponents of these methods that they will reduce the effort required to create a justified land-management strategy for a large region by reducing the area which will need to be surveyed, and hence reduce the expenditure. Again, Kvamme describes the motivation for the development of the techniques well:

"If powerful resource location models can be developed then cultural resource managers could use them as planning tools to guide development and land disturbing activities around predicted archaeologically sensitive regions. This planning potential of predictive models can itself represent significant cost savings for governmental agencies." (Kvamme 1990a:289)

It is certainly the case that the countries most directly equated with GIS-based predictive modelling (the US, Canada and the Netherlands) are those facing either a combination of very large study areas and relatively few archaeologists and/or very vigorous, ongoing programmes of redevelopment (*e.g.* Kamermans 1999, Dalla Bona 2000).

Before we go on to discuss predictive modelling in detail, it is important to acknowledge that of all of the applications of GIS practiced by archaeologists, predictive modelling has proven the most methodologically and theoretically controversial. Some archaeologists feel that it should not be undertaken at all, either because it does not work well enough to be useful or because it is theoretically undesirable to explain cultural phenomena in this way. We will expand on this later in the chapter, but first it is necessary to review the various methodologies that have been used for predictive modelling.

8.1 DEDUCTIVE AND INDUCTIVE APPROACHES

It is important to realise that the term *predictive modelling* does not refer to a single method, but essentially describes a wide range of approaches to one specific problem—namely will a location contain archaeological material or not? Predictive models can be based on two different sources of information:

1. theories about the spatial distribution of archaeological material;
2. empirical observations of the archaeological record.

When a model is based entirely on theory then the model is referred to as a *deductive* or *theory-driven* model. If the model is based entirely on observations it is referred to as an *inductive* or *data-driven* model, sometimes referred to as a *correlative* model. Conceptually, it should be possible to create a model based either on pure induction, or on pure deduction: for example we may theorise a relationship between site location and environmental resources and then implement this relationship as a predictor in a GIS without any recourse to the actual locations of the sites. Alternatively, it may be possible to derive a mechanical method for using the environmental characteristics of a group of known sites to derive such a relationship without recourse to a hypothesis about the function or meaning of the sites—in effect projecting a known pattern into an unknown area.

It should be recognised that such a distinction between data and theory is not universally recognised, and most archaeologists accept that the two are not independent—data is collected within a theoretical context, and so may be regarded as theory-laden, while theories are generally based to some extent on empirical observations. Although it is impossible in any practical sense to implement a predictive modelling method that is based entirely on either of these tactics, it is probably useful to maintain the distinction between the two approaches for the purposes of discussion.

8.2 INPUTS AND OUTPUTS

Predictive modelling can be reduced, in essence, to the production of one or more outputs from one or more inputs through the use of a specified rule. In the case of archaeological predictive modelling, the outputs are always archaeological in nature while the inputs are generally not.

Unit of analysis

It should be noted that the 'unit of currency' of archaeological predictive models is the landscape location, not the archaeological site—in a raster system this means that the object of the model is the grid cell of specified resolution. The model then works on this unit, to assess the output from the inputs. From this it should be obvious that the archaeological site is *not* central to the construction of these models: instead the unit of landscape (most often a grid-cell of a raster-GIS) is the basic unit for which attributes are recorded (Kuna 2000). These attributes are archaeological, and are most usually the presence/absence of one or more classes of archaeological site. Alternatively, the attributes of the landscape unit might be more sophisticated, for example: the density of sites; the number of artefacts per unit area; or even the type of sites (as a nominal variable).

Inputs to models

Input to predictive models is usually in the form of non-archaeological information about the landscape locations. Various inputs have been used, including the following.

Spatial parameters

In cases where sites are known to have spatial relationships—for example if they tend to cluster or to be noticeably dispersed—this information can then be used as a predictor in the model. For example, if it is assumed that a particular type of site is likely to form a clustered distribution, then the likelihood of an undiscovered site occurring in any given location will decrease with distance from known sites. This means that, when applied to archaeological materials, *spatial interpolators* can in some sense be regarded as predictive models.

Physical environment characteristics

These are the most easily obtained inputs in GIS contexts, and therefore the most widely used inputs to models. Which variables are selected should be determined by the knowledge that they are related to the location of the sites in question. This may be determined either by the experience of a particular archaeologist or more formally by a statistical analysis of existing sites. A huge variety of characteristics may be used as predictors, and those that have been employed include elevation, landform derivatives such as slope, aspect, indices of ridge/drainage, local relief, geological and soil data, nominal classifications of land class (such as 'canyon', 'plain', 'rim' *etc.*) and distances to resources such as water or raw material sources. In cases where the vegetation is considered not to have changed significantly, vegetation classes may also be used. It is important to bear in mind that whatever inputs are used they are representative of the *modern* environment rather than that of the particular period of the past you may be interested in.

Economic measures

These are one step removed from the environment, in the sense that they require more theoretical inference. These may include the *productive capacity* of the land, inferred from soil and other resources and from palaeoeconomic theory, or the suitability of different places for particular crops.

Cultural features

These are more difficult to obtain reliable or complete quantitative measures of, and are therefore more difficult to include as inputs to formal models. Indeed, in an important review of predictive modelling Kohler has argued that one of the key reasons archaeologists have avoided using them in predictive models is that they simply do not know *how* to use them (Kohler 1988:20). In some cases, the cultural features of a landscape can be of use in predicting the location of sites. For example, it may be that sites occur in close proximity to road networks, central places or focal (significant) points in the natural landscape. The relationship of sites to factors such as these requires yet more 'archaeological' input to the model.

Outputs from models

The output from the model may be very simple or fairly complex, and may take a variety of specific forms. For example:

• *Presence/absence of site.* This is the most common type of output required from a predictive model. These types of models usually represent one specific site type with general characteristics, usually defined by function or by chronology—for example a hunting camp. Output in this case is binary.

• *Site class.* This can be output from more sophisticated models. In this case a decision is made about each location, which places it in one of several site classes or a special class of 'no site'—hunting camp; processing site; base camp. Output from these models is categorical.

• *Densities of sites/artefacts.* In situations where the resolution is large enough so that several sites or artefacts tend to occur in each location, then density can be considered as the output variable of the model. In this case the output may be of ordinal or real number data.

• *Site significance.* Occasionally resource management applications have attempted to use prediction to classify landscapes according to the perceived importance of the archaeological remains that may be present. In this case the output may then be of any type, depending on how 'significance' is defined.

• *Site probability.* One particularly popular type of model produces as its output an indication of the probability of a site occurring at each location in the landscape. In these instances the output is usually a real number, either ranked between 1 and 0 or expressed as a percentage.

8.3 RULE-BASED APPROACHES

The simplest way to construct a predictive model is to generate a *decision rule*, which describes the conditions under which a particular location is likely to contain an undiscovered site. This rule can be based on one or more predictor variables, and can be derived either entirely from theoretical hypothesis (deductive) or from observations of an existing sample of sites (inductive/correlative).

Deductive rule approaches using map algebra

If a simple decision rule is used, then the outcome of each implementation of the rule will produce a yes or no answer: yes if the rule determines that a site is likely to be present, or no if it is not. More sophisticated decision rules may be able to produce a variety of outcomes, representing different classes of archaeological outcome—typically different densities of remains, or different types of sites.

The inputs to a decision rule are generally a subset of the non-archaeological characteristics of the region in question. Thus the first assumption of the decision rule is that the locations of archaeological sites are related in some way to the non-archaeological characteristics of the region. In many cases this can be justified by observation of an existing set of sites, but occasionally no sites, or only a very small sample of existing sites are available. In these cases a sound archaeological case should be established for the assumption.

At their simplest, decision rules can be regarded as functions of the variables (themes) which are used as inputs. This can be described as in Equation 8.1, where M represents the case for the model indicating the presence of a site, and M' indicating the opposite (that no site occurs).

$$\{f(x_1, x_2, x_3, ...) > 0\} = M$$
$$\{f(x_1, x_2, x_3, ...) \leq 0\} = M' \qquad (8.1)$$

Because this function produces a binary, mutually exclusive classification using more than one input, this type of function is called a *multivariate discriminant function*.

Another approach is to use set notation to define M as the *intersection* of a variety of variables, in which case M' is the complement of that action. A typical example of such a decision rule might be the hypothesis that a particular class of site will tend to occur in flat areas (slope of less than 10 degrees), not too far from a source of fresh water (less than 1km), on a particular type of soil (class A) and in south facing aspects. This can be expressed as a decision rule, using set notation as follows:

$$M = (Slope < 10) \cap (Distance < 1km) \cap (Soil = A) \cap (Aspect = South)$$
$$M' = (Slope \geq 10) \cup (Distance \geq 1km) \cup (Soil \neq A) \cup (Aspect \neq South) \qquad (8.2)$$

Some GIS will allow rules to be specified in their entirety in some high-level language, examples of this include the MapInfo 'mapbasic' language, GRASS 'r.mapcalc', ArcView Avenue or Arc Info AML. Simpler systems require that the decision rule be implemented in a series of steps using single operators.

Inductive methods of obtaining simple rules

Instead of defining the map algebra rules for a model deductively, it is also possible to allow the characteristics of an existing set of data to determine the rules. The simplest way that this can be done is by classifying the predictors into a fixed number of classes, and then observing the distribution of the sites in each of the predictor variables.

To give a hypothetical example, we may have a series of site locations located by field walking in a study area that can be interpreted on the basis of material culture as representing early prehistoric agricultural settlements. For the purposes of developing a predictive model to highlight potential locations for such sites outside of the fieldwalked area, we may want to look at the types of soil favoured by these early farmers. To do this we would typically take a modern soil map of the study zone and reclassify it into three classes (good, moderate, poor) on the basis of its suitability for cultivation using a simple ard plough.

From this an *expected* number of sites can be calculated for each class of each predictor (the total number of sites multiplied by the proportion of the total area occupied by that class). Each class of each variable can then be weighted so that where sites are over-represented they have a high weighting, while where they are under-represented they are given a lower weighting.

The precise allocation of weights to the classes in each predictor can be varied: some archaeologists have used the ratio of expected sites to observed sites directly as a weighting, others have opted to assign integer weightings which can be positive for 'favourable' areas, and negative for 'unfavourable' areas.

At the same time, a significance test (chi-squared or similar) can be undertaken to ascertain whether the proposed predictor really can be held to be associated with the distribution of archaeological material at a given confidence interval. If it cannot, then the variable should probably not be used in the model. If a chi-squared test is used, then the phi-squared measure of association can also be calculated and each input variable can also be weighted according to the strength of the association between predictor and data (Table 8.1).

However the class weights are assigned, the individual predictors are then

Table 8.1 One method of calculating 'weights' for an input theme.

Distance from water	0-499m	500-999m	1000m +	Total
Proportion of area	60%	35%	5%	100%
Number of sites	15	45	10	70
Expected	42	24.5	10	
Weighting	*0.35*	*1.8*	*0.3*	

reclassified into 'site favourability' maps, and these are then added together to produce an overall prediction.

Although conceptually and statistically quite simple this approach takes no account, for example, of the possible correlations *between* input variables. That this can be highly significant has been shown by, amongst others, (Cumming 1997) in his evaluation of the extent to which Sites and Monuments Records (SMRs) in the UK are an accurate reflection of the underlying archaeological record.

8.4 REGRESSION-BASED APPROACHES

Although the deductive and inductive approaches discussed above represent an adequate method of generating predictions, such models do leave a number of important questions unanswered. For example:

- To what extent does each predictor affect the model?
- How might ordinal or continuous archaeological variables (such as site or artefact densities) be predicted?
- How good a fit is the model to the observed distribution of the sites?

These questions are exactly the kinds that are best answered by regression analysis, where the effect of one variable on another can be modelled by a mathematical equation. Several regression techniques can be used to produce archaeological predictive models, the most common being *linear multiple regression* and *logistic multiple regression*. These are appropriate for different kinds of modelling problems and each has particular strengths and weaknesses.

In each instance, the cases in the regression are land parcels; the dependent variable is some known archaeological outcome (amount of archaeological material or presence/absence of archaeological sites) while the independent variables are the supposed predictors of the archaeological outcome.

In the case of GIS-based studies, this raises a problem. Often we wish to perform a regression on the archaeological material in a region, represented as a raster layer. Each cell in the raster represents a land parcel and there may be layers for archaeological outcome (which we want to predict) as well as appropriate 'predictors' for use as independent variables. The problem is that raster cells which contain no record of archaeological material do not represent *zero* values unless they are areas which have been surveyed and shown to be free of the archaeology in which we are interested. Instead, they usually represent locations for which we have *no data* about the archaeological outcome, and are therefore not suitable for use as cases.

Where extensive surveys have been done, then there is not really a problem. The land parcels within the surveyed areas can be used as the cases for the regression, and the resulting equation (relating predictors to archaeological outcome) can then be applied to areas for which no archaeological data is recorded.

However, where models are trying to 'fill in the gaps' within existing site distributions then there are no land parcels with a 'no site' value, and it may be necessary to assume that the vast majority of the non-archaeology landscape cells will contain no archaeology, and select a random number of locations to use as 'no archaeology' cases.

Linear multiple regression

Linear multiple regression techniques can be used where the archaeological outcome we are interested in is an ordinal level variable or higher—such as the size of something or the number of artefacts per hectare. The procedure is fairly straightforward, and involves generating a table of the archaeological outcome of raster cells with known values together with the values of those raster cells in all proposed predictor cells. The regression can then be run and, although a variety of algorithms may be chosen, all produce an equation of the form:

$$y = a + b_1x_1 + b_2x_2 + b_3x_3 + ... + b_kx_k \tag{8.3}$$

where y is the dependent (archaeological) variable, a is a constant and $b_1 .. b_k$ are all the independent variables (predictors). As well as this equation—which can be used to predict the values of the archaeological variable in unknown locations through the use of map-algebra—multiple regression also produces the correlation coefficient r^2 for the model. This indicates the proportion of the variation within the archaeological variable that is 'explained' by the equation. It also produces partial r^2 values for each of the predictors, which can be particularly archaeologically interesting and useful as they suggest the contribution that each of the predictors has made to the model.

Problems with linear multiple regression

Although linear multiple regression can be a useful approach to predictive modelling, it does have several restrictions, which can make it inappropriate in many situations. Firstly, it requires that the archaeological outcome in which we are interested can be measured on an ordinal scale or higher. This is sometimes the case, but frequently we are simply interested in site presence or absence, or in classes of site.

The independent variables also have to be measured on an ordinal scale or higher so it is difficult to include choropleth (nominal) overlays in the model. Spatial data is frequently nominal in nature (*e.g.* soil, geology) rather than ordinal and, although it is possible to break these variables down into 'dummy' variables for use in multiple regression, their use is difficult. As well as this, linear multiple regression performs badly when the independent variables are not normally distributed, or when the relationship between predictor and predicted is not linear (for more details of both bivariate and multivariate regression analysis see *e.g.* Shennan 1997 Chapters 8, 9 and 10). This is often the case with the kinds of spatial variables which we may want to use as predictors (see Figure 8.1), meaning that the variables may need to be transformed by the use of some operator into a distribution that more closely approximates a normal one prior to their use in the regression.

Logistic multiple regression

An alternative to linear multiple regression is the use of probability models to predict archaeological outcome. These are a family of models that produce

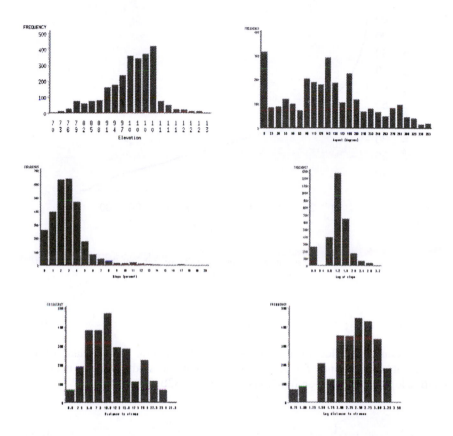

Figure 8.1 Frequency distributions of elevation (top left), aspect (top right), slope (centre left), log of slope (centre right), distance from streams (bottom left), and log of distance to streams (bottom right).

estimates not of the absolute value of a variable but of the *probability* of a particular outcome. The most widely used method in archaeology has been *logistic regression analysis*.

The main advantages of logistic regression over other forms of regression are that it makes fewer assumptions about the form of the independent variables—the level of the variables (including the archaeological dependent variable) can be ordinal level or higher—and the result can be interpreted as the probability of a particular archaeological outcome. When the outcome is site presence/absence, then the result can be interpreted as the probability of site presence, given the values of the independent variables.

In practice, logistic regression procedures operate in the same kinds of way as other forms of multiple regression. The scores of the surveyed landscape units are recorded in a table, showing both the archaeological outcome and the values of the other predictors. The logistic procedure is then run, producing an equation of the form:

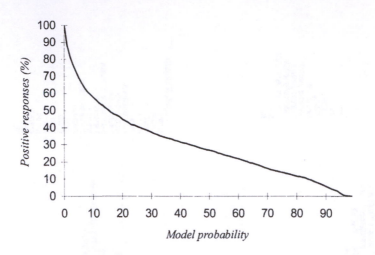

Figure 8.2 The proportion of the observations predicted as a site for each probability result from a logistic regression equation.

$$L = a + b_1x_1 + b_2x_2 + b_3x_3 + ... + b_kx_k \qquad (8.4)$$

The calculated value of L can be used to generate an estimate of the probability of site presence for all locations within the study area by using the cumulative logistic distribution function (Kvamme 1983b: 371):

$$p = \frac{e^L}{1+e^L} = \frac{1}{1+e^{(1-L)}} \qquad (8.5)$$

Although the map algebra required to apply the model is a little more complicated, the procedure has proved to be a very effective and flexible method of generating predictive probability models.

Bias towards successful prediction of nonsites

One problem with logistic regression is that the value of L is adversely affected by situations in which the number of positive responses on the dependent variable (sites present) is significantly higher than negative ones (site absence) or vice versa. In this situation, the resulting estimate of L will be biased towards prediction of the larger class. In archaeology, the larger class is usually site absence so this is often an undesirable effect.

Two possible mitigation strategies are possible: the first is simply to select a random sample of the negative responses so that there is the same number of positive and negative responses, although this involves discarding an often large

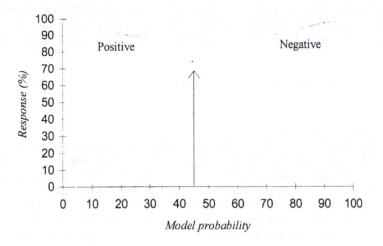

Figure 8.3 Identifying the cut-off point for a logistic regression model.

number of valid observations. The second approach is to attempt to correct for the sample size bias after running the model, which can be done by adjusting the intercept parameter (the constant a) used to estimate L by the natural log of the ratio of the sample sizes (Kvamme 1983b).

However, a more significant problem with logistic regression is that it does not produce the same easily interpretable correlation coefficients as multiple regression. As a result, it is far more difficult to judge how effective the model is in explaining the archaeological outcome. A variety of methods can be used to estimate the effectiveness of the model, most common of which is withholding a subset of the observations from the analysis, and comparing the results of the regression with the withheld observations.

Optimising logistic regression models

A useful characteristic of logistic regression models is that where site presence/absence is the dependent variable, they produce a probability of site presence as their output.

The 'optimum' point of the model so far as providing correct predictions of site presence/absence can be identified by comparing the known archaeological outcome (from the original data) with the predicted outcome (from the equation). By selecting different values of p (the model probability) as the value for predicting site presence, the proportion of the original observations that are predicted as site presences can be altered. This is because as p is increased, the number of cases that are predicted as archaeological sites is decreased (see Figure 8.2).

It is also possible to graph the proportion of the original observations that are *correctly* classed as sites against those *correctly* classed as nonsites for each value of p. This produces a graph such as that shown in Figure 8.3, where the line for proportion of correct nonsite predictions crosses the line for proportion of correct site predictions at some value of *p*. This point is the point at which the model gets the answer right for the same percentage of sites as it does for nonsites, and is the value of *p* often regarded as representing the optimum performance of the model.

8.5 AN EXAMPLE: PREDICTIVE MODELLING IN ACTION

By now you should have a good appreciation of the variety of techniques and methodological approaches that are available for predictive modelling. Undoubtedly the best way to illustrate some of the issues outlined above is through a worked case study and the example we have chosen relates to a recently published exercise carried out by Westcott and Kuiper (2000). This study seeks to create an inductive model of prehistoric site distributions in a large, unsurveyed coastal area of the Upper Chesapeake Bay. The problem facing the archaeologists centred upon a study zone known as the Aberdeen Proving Ground (APG)—a 30,400-hectare area that had been owned and utilised by the army for much of the 20th century. Due to a combination of restrictions, unexploded ordnance and marshy conditions, only 1% of the overall area had witnessed any form of structured archaeological survey, with only 46 prehistoric sites known from the entire zone. Hence the need for a predictive model to facilitate the effective management of the archaeological resource.

Ideally the formulation of the decision rule for the predictive model for the APG would be based upon the results of a controlled, intensive survey within the actual APG zone. However, for the reasons outlined above this was not possible. As a result the researchers decided to use what they termed 'available data' as a proxy. This comprised careful study of a database of 572 prehistoric sites located in areas of the Upper Chesapeake Bay that most closely resembled the environment of the APG. In an important deductive step these sites were broken down into two distinct classes: shell-middens and lithic scatters, which were assumed to have been located with respect to different, although overlapping, sets of variables. For each of these sites a range of locational environmental factors (*e.g.* proximity to water, soil, soil drainage, topographic setting, slope, aspect *etc.*) were recorded along with a comparative 'background' sample of 500 random locations taken from within the APG.

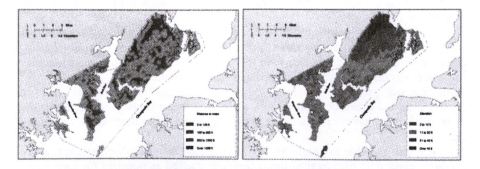

Figure 8.4 Examples of the environmental variables selected as inputs for the final predictive model: distance to water (left) and elevation (right). From Westcott and Kuiper (2000:63-65). Reproduced with permission.

To decide upon which variables to use in the construction of the predictive model an inductive procedure was followed based upon the generation of a series of frequency tables for the site and random background locations with respect to each of the environmental variables and combinations of variables. This enabled a number of variables to be eliminated—for example in the case of soil drainage the percentage of site locations occurring on well drained soils was not significantly different to the percentage of well-drained soils occurring in the background data.

In addition, aspect was inconsistently recorded on the site datasheets and seen to be closely correlated with the particular shoreline (east or west). As a result of this exercise the following set of four core environmental predictors were identified (Figure 8.4):

1. distance to water (fresh, brackish);
2. water type (*e.g.* river, creek, swamp);
3. elevation range;
4. topographic setting (*e.g.* floodplain, terrace, hilltop, bluff).

Although logistic and linear regression techniques were considered, it was acknowledged that the 'available data' approach taken did not meet the statistical assumptions of these models. Instead, the weightings for particular combinations of classes possible between the four predictors were determined by comparing site locations from the proxy data set to the various combinations (see Table 8.2).

A high potential designation was then assigned to areas where sites occurred in any given unique combination of the four variables over 20% of the time; medium potential for sites occurring between 6.25 and 20% of the time; and low potential for the remainder. Construction of the predictive model then involved combining the four variables and allocating each cell its appropriate potential classification dependent upon the unique combination of variables. The resultant shell-midden predictive model is shown in Figure 8.5.

The efficacy of the final model (shown in Figure 8.5) was initially evaluated by comparing the known 46 prehistoric site locations within the APG to it. Looking to the shell-middens of the 13 known in the study area, 12 fell within the high potential zone and 1 in the low potential. A more formal assessment of the overall performance was undertaken using Kvamme's simple gain statistic:

Table 8.2 The frequency table generated for the shell midden sites. From Westcott and Kuiper (2000:67). Reproduced with permission.

Distance to water (ft)	Water type	Elevation (ft)	Topography	Frequency	Percentage	Cumulative frequency	Cumulative percentage
0-500	Brackish	≤ 20	Terrace/Bluff	75	34.7	75	34.7
0-500	Brackish	≤ 20	Floodplain/Flat	81	37.5	156	72.2
0-500	Brackish	> 20	Terrace/Bluff	14	6.5	170	78.7
0-500	Brackish	> 20	Floodplain/Flat	2	0.9	172	79.6
0-500	Fresh	≤ 20	Terrace/Bluff	24	11.1	196	90.7
0-500	Fresh	≤ 20	Floodplain/Flat	10	4.6	206	95.4
0-500	Fresh	> 20	Terrace/Bluff	4	1.9	210	97.2
> 500	Brackish	≤ 20	Terrace/Bluff	4	1.9	214	99.1
> 500	Brackish	≤ 20	Terrace/Bluff	2	0.9	216	100.0

Gain = 1 – (% of total area covered by model/% of total sites within model area)   (8.5)

This is predicated upon the assumption that if the high potential area is small relative to the overall study area and that if the number of sites found within it is large with respect to the total for the study area then we have a fairly good model (Kvamme 1988:329). The closer the value of gain is to 1 the better the models predictive utility. With a calculated value 0.82 for high potential areas and 0.80 for medium predictive areas the model's performance was deemed to be good (Westcott and Kuiper 2000:69).

8.6 METHODOLOGICAL ISSUES IN PREDICTIVE MODELLING

Despite the apparent methodological sophistication of methods such as logistic regression, there are a number of criticisms that can be aimed at the construction of predictive models. Woodman and Woodward (in press) have usefully reviewed the statistical and methodological basis of predictive modelling in archaeology and noted several outstanding problems.

They draw attention to the fact that 'case-control' studies are normally used in predictive modelling. Instead of sampling the landscape and then calculating the percentage of sampled points that turn out to be sites, it is quite usual to gather a sample of sites and a sample of nonsites. This means that all estimates of the probability of site presence must be relative estimates rather than absolute: it is not possible to use these studies to predict that a location has x% chance of containing a site and another location a y% chance, although it may be possible to state that

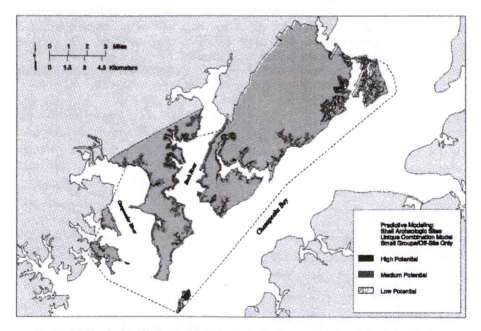

Figure 8.5 The final predictive model produced for the shell-midden sites. Westcott and Kuiper (2000:68). Reproduced with permission.

location *x* is twice as likely as location *y*.

Woodman and Woodward also criticise the absence of attention in the predictive modelling literature to the data requirements. Procedures such as logistic regression assume a linear relationship between dependent and independent variables, an assumption that is rarely tested. Similarly, they point out that very little attention is given to the nature of the interactions between the independent variables, which can take the form of correlation, confounding or interaction.

Ebert (2000) is also highly critical of inductive predictive modelling methods, arguing that methods used to test the accuracy of models by comparing results that have been obtained with one part of a sample against another (withheld) part of the sample constitute little more than "a grossly inefficient way to determine if there is inhomogeneity in one's data" (ibid:133). He is also critical of the way that maps are translated directly into variables, and of the accuracy of published models, concluding that:

"Inductive predictive modelling ... is not going to get any more accurate than it is right now. It focuses on the wrong units of analysis, sites rather than systems, and attempts to relate their locations to 'environmental variables' which not only are probably not variables at all, but cannot be warranted by any theoretical argument to be effective predictors of the locations of components of systems across landscapes." *(*ibid:133)

8.7 THE PREDICTION PREDICAMENT: THEORETICAL DIFFERENCES OF OPINION

"Most archaeological predictive models rest on two fundamental assumptions: first, that the settlement choices made by ancient peoples were strongly influenced or conditioned by characteristics of the natural environment; second that the environmental factors that directly influenced these choices are portrayed, at least indirectly, in modern maps of environmental variation across the area of interest." (Warren and Asch 2000:6)

As we said earlier in this chapter, the development and application of predictive models of archaeological site locations has been controversial. We might go so far as to say that predictive modelling of archaeological site patterns is the most controversial application of GIS within archaeology and although this volume is not intended to contribute or extend this debate, the reader will need to be aware of the essential differences of opinion that exist within the discipline (see *e.g.* Gaffney and van Leusen 1995, Kvamme 1997, Wheatley 1998, Kuna 2000).

Central to the critique of predictive models has been the accusation that the entire methodology embodies a theory of archaeology that can be categorised as *environmental determinism*. The debate about environmental determinism and predictive modelling has been vigorous. Some advocates of predictive modelling claim that the utility of models to identify patterns can be isolated from their use as interpretative tools. As van Leusen puts it:

"Statistics can be used to describe patterning in archaeological datasets in a rigorous manner without reference to the cause(s) of those patterns, and if the extrapolation of those patterns yields predictions that are useful in CRM ... the method is validated." (Gaffney and van Leusen 1995:379)

Critics of predictive modelling, as currently practiced, have tended to argue that the description of patterns, and their modelling for whatever purpose cannot be separated from the process of explanation in this way. Even if it could be, then the correlation of static archaeological settlement patterns with environmental variables constitutes a worthless exercise. The argument that inductive modelling can therefore be separated from deductive, explanatory modelling has been described by Ebert as a product of the "sheer indolence among those who think we can 'stop' at 'inductive predictive modeling' " (Ebert 2000:130). He goes on to argue that:

"Predictive modeling will be transformed into a worthwhile adjunct to archaeology and archaeological thinking only by the formulation of a body of explanatory propositions linking contemporary correlations with the past. In other words it is productive explanatory thought, and not computers, that can potentially raise predictive modeling above an anecdotal level." (ibid)

Other defenders of locational models have objected to the use of the term 'environmental determinism' to describe the approach, arguing that the prevalence of environmental variables in archaeological predictive models is simply a product of their greater availability over 'cultural variables' and so should not be taken as an indication of theoretical orientation. To quote Kohler:

"Given the subtleties and especially the fluidity of the socio-political environment, is it any wonder that archaeologists have chosen to concentrate on those relatively stable, distorting factors of the environment for locational prediction?" (Kohler 1988:21)

Moreover, some (Kvamme 1997:2) have objected on the grounds that the term is an inappropriate—even offensive—label that associates predictive modellers with late 19th and early 20th century human geographers such as Rahtzel (1882), whose ideas were later widely discredited.

8.8 CONCLUSIONS

We have attempted in this chapter to introduce some of the methods central to a large field of research. We have tried to indicate some of the methods of (mainly inductive) modelling that have been applied to archaeological situations, devoting greatest attention to the method that is currently most popular, logistic regression. We have also tried to draw attention to some of the unresolved methodological issues surrounding the statistical methods and to some of the theoretical concerns that have caused so much debate in the archaeological GIS literature.

Inevitably, constraints of space have prevented us from exploring any but a small subset of the range of possible methods for the generation of models and the interested reader will want to follow up many of the references included here, particularly the papers published in Judge and Sebastian (1988) and, more recently, Westcott and Brandon (2000).

The perceptive reader will also have understood that predictive modelling is not a field in which we have any programmatic interest ourselves. While we have attempted to write a balanced and reasonably complete account of archaeological predictive modelling, it is unlikely that we have been successful in completely disengaging from an academic debate in which we have been quite active protagonists. Both authors have published criticisms of predictive modelling

(Wheatley 1993, Gillings and Goodrick 1996) and would subscribe to the position that it is a field with significant unresolved methodological and, more significantly, theoretical problems.

Some of these we have introduced above, but a theoretical debate such as this is difficult to synthesise without reducing it to the level of caricature. We therefore hope that archaeologists who want to take a more active interest in this field will follow up references to the debates and discussions that have been published in the archaeological literature (notably Kohler 1988, Gaffney *et al.* 1995, Kvamme 1997, Wheatley 1998, Church *et al.* 2000, Ebert 2000) and formulate an independent opinion as to whether Predictive Modelling has a future in archaeology.

CHAPTER NINE
Trend surface and interpolation

"Everything is related to everything else, but near things are more related than distant things." (Tobler 1970)

Interpolation refers to the process of making mathematical guesses about the values of a variable from an incomplete set of those values. Spatial interpolators use the spatial characteristics of a variable, and the observed values of that variable to make a guess as to what the value of the variable is in other (unobserved) locations. For example, if a regularly spaced auger survey revealed an easily identifiable horizon, then it may be desirable to try to produce a contour map of that deposit from the points of the survey. In this case a spatial interpolator would have to be used to predict the values of the variable (in this case the topographic height of deposit X) at all the unsampled locations.

Chapter 5 has already introduced one specific type of interpolation problem: that of estimating topographic height from contours. However, we noted then that interpolation has both a far wider range of applications than merely to elevation data, and also encompasses a very wide range of methods. In this chapter we will review some of the main methods for interpolating continuous values from point data, and comment on their applications to archaeological situations.

There are a wide variety of procedures for spatial interpolation, each with different characteristics. It is important to note that the choice of interpolation procedure should be dependent on the nature of the problem. In addition, a sound understanding of how each interpolator works is essential if they are to be used effectively.

It is possible to use the spatial allocation methods discussed in Chapter 7 as crude spatial interpolators. Point data can be used to generate a Voronoi tessellation, and the values of the points allocated to the polygon defined by it (the effect of this can be seen in Figure 9.7, top left). The major disadvantage of this is that it does not provide a continuous result. Many archaeological variables, such as artefact density or dates, do not conform to finite boundaries and although there are methods for interpolating from area data to continuous variables—see *e.g.* Burrough (1986)—these are rarely, if ever, used in archaeology.

The remainder of this chapter provides a brief overview of some of the methods available for interpolation of points into density estimates and of ratio or higher scale observations into continuous surfaces. We have already discussed one continuous surface, topographic height, but other examples could include any archaeological or environmental value that can be measured on an ordinal scale or higher and for which we would like to estimate values at unobserved locations. Examples might include artefact density, rainfall, ratio of deciduous to coniferous species and many more.

These are illustrated with images showing the results of applying different procedures to a dataset of 176 points for which there is a measurement of some continuous variable (originally elevation in metres, but we will imagine that it could be anything because some procedures would not be appropriate for elevation data), and for which there is also a record of presence/absence of some diagnostic characteristic at that location. For clarity, the same dataset has been used throughout the chapter, although in places it is treated simply as a set of points.

Some of this chapter resorts to a rather formal, mathematical notation to describe the methods, although we have omitted considerable detail for the sake of brevity. We realise that we have not been particularly consistent in the level of mathematical detail included, but chose to give quite a lot of the formal notation for some procedures in preference to providing a little maths for all of them. A reader who does not have enough familiarity to follow the mathematical notation should still be able to follow the main differences between these approaches described in the text, and if more detail is required then a more specific text should be consulted. Bailey and Gatrell (1995) covers the vast majority of the methods described in this chapter with particular emphasis on geostatistical methods. This text also has the considerable advantage that it includes exercises, datasets and software allowing the reader to gain hands-on experience of the procedures without needing access to expensive hardware and software. Alternatively (or additionally) Isaaks and Srivastava (1989) provide a comprehensive and easily followed introduction to applied geostatistics that would form the ideal starting point for further study.

9.1 CHARACTERISTICS OF INTERPOLATORS

One characteristic of an interpolator is whether or not it requires that the result passes through the observed data points. Procedures that do are *exact interpolators* as opposed to *approximators,* in which the result may be different at the sampled locations to the observed values of the samples.

Another important characteristic is that some procedures produce results whose derivatives are continuous, while others produce surfaces whose derivatives are discontinuous, in other words the surface is allowed to change slope abruptly.

Most importantly, some interpolators are *constrained*, so that the range of values that may be interpolated is restricted, while others are *unconstrained* where the interpolated values might theoretically take on any value. In some instances, particularly in situations where you are predicting things that cannot be negative, the use of unconstrained procedures should obviously be avoided.

We might also classify interpolation approaches by the extent to which they can be thought of as *global* or *local* procedures. Global approaches derive a single function that is mapped across the whole region, while local interpolators break up the surface into smaller areas and derive functions to describe those areas. Global approaches usually produce a smoother surface and, because all the values contribute to the entire surface, a change in one input value affects the whole surface. Local interpolators may generate a surface that is less smooth, but alterations to input values only affect the result within a particular smaller region.

Although we pay little specific attention to it in this chapter, readers should also be aware that some of the procedures could be influenced by *edge effects*. These occur in situations where estimations rely on a neighbourhood or region surrounding a point, or on the number of points close to an unknown location. In both of these cases the estimation will not be possible close to the edge of the study area and, although various compensations are used to allow for this, it may still produce unpredictable results near the edge of the study area.

9.2 POINT DATA

There are many occasions when archaeologists have spatial observations that are not measured on a numerical scale at all. Instead, we may have points that represent observations that something is merely present at a particular location. Examples of this kind of data include sites at a regional scale or artefacts within smaller analytical units. Storing this kind of data within a spatial database is unproblematic, as it consists of simple point entities to which we may also attach other attributes such as size or period (for sites) or object type and classification (for artefacts).

Archaeologists have often used methods of 'contouring' this type of data into density or other continuous surfaces. Although several procedures exist to do this, careful thought needs to be given to whether or not it is an appropriate or meaningful thing to do: it *may* be useful to 'interpolate' from the observations in order to estimate the density of similar artefacts or sites that might be expected to occur at unsurveyed locations. On the other hand, the interpolation of artefact densities within a wholly surveyed area into a continuous product is of entirely dubious utility. We have seen this done—for example using the densities of artefacts within features to create a surface that includes excavated but empty parts of a site—but would caution strongly against it unless there are strong analytical reasons to do so. The 'interpolated' values do not represent an estimate of anything real, because we already *know* where all the artefacts actually are and we are, essentially, generating a wholly artificial surface. Very often, archaeologists would be better advised to concentrate on the presentation of the original data values using choropleth maps, symbols or charts (to show proportions) rather than creating dubious 'interpolated' continuous products. With that proviso, we can note that there are several methods of converting presence/absence data stored as points into an ordinal or continuous product.

The most straightforward method is to generate a grid in which the attribute of the grid cells is the density of the points that occur within them. This is essentially the same as the quadrat approach to point pattern analysis that we discussed in Chapter 6 and presents the same problems, notably of choosing an appropriate grid/quadrat size. It requires a grid resolution sufficient to ensure that many grid cells contain several points, but not so large that the subtle variations in artefact densities are obscured by aggregation. Different grid cell sizes may appear to produce rather different patterns, so it is important to pay careful attention to this, ideally generating densities with a number of different cell sizes. This method is also not optimal because the shape of the grids/quadrats, almost always squares, can affect the density surface.

This approach has been followed in many archaeological examples—see *e.g.* Ebert, Camilli and Berman (1996:32) for an example using densities of lithic flakes and angular debris) but its popularity may derive more from the intuitive familiarity that fieldworkers have with the way that it works than any inherent methodological advantages. It is, after all, a computational expression of what archaeologists have done for many years with fieldwalking data and paper—see *e.g.* papers in Haselgrove, Millett and Smith (1985).

Another approach is to generate a circular area of radius r centred on each point and give it, as an attribute, the density derived from its source point (in other words $1/\pi r^2$ where r is the radius of the circle). These circles can be added together to give what is sometimes referred to as a *simple density* operator. As the radius for the values increases, the density surface becomes more and more generalised. Figure 9.1 shows clearly how this affects a given set of points. It will be apparent that the density surface produced can be as local or as general as we wish and care should therefore be taken to select an appropriate radius for the task in hand. Although this takes account of the arbitrary nature of square sample units, it unfortunately should also be clear that the greater the search radius becomes the more influenced the resulting surface will be by *edge effects* although it is possible to adjust the initial density values of those circles which are only partially within the area. The bottom two images of Figure 9.1 were calculated without making any adjustments in this way and are therefore significantly influenced by the edges of the study area.

A related approach is the use of Kernel Density Estimates (KDE). These have recently been introduced to archaeology—see *e.g.* Baxter, Beardah and Wright (1995), Beardah (1999)—although not specifically for spatial interpolation. KDE operate in a similar manner to simple density estimates except that the circle centred at each point is replaced by a density function called the *kernel*. In simple terms, the kernel describes a 'bump' at each data point and these can be added together to arrive at a density estimate. The kernel itself can be any bivariate probability density function which is symmetric about the origin and which can be controlled with some bandwidth parameter. Just as with the radius of the circles, the bandwidth parameter (sometimes parameters) will determine the smoothness and other characteristics of the surface. Further details have been omitted here for the sake of brevity, but Bailey and Gatrell (1995:84-88) would make a good starting point for further reading on the use of KDE as a spatial interpolator.

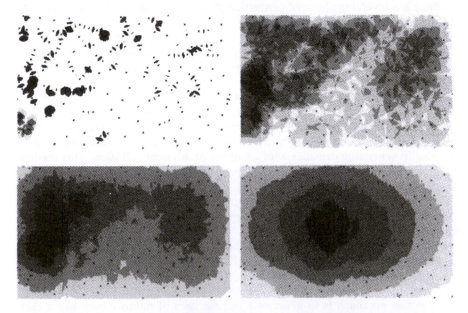

Figure 9.1 Simple density estimates for a point data set using values of 150m (top left), 500m (top right), 1000m (bottom left), and 2500m (bottom right) for the radius of the circles. Grey scale varies. Generated with ArcView.

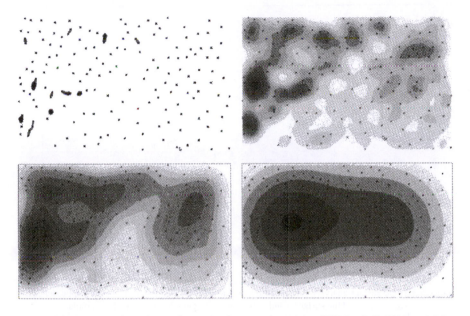

Figure 9.2 Kernel density estimates for a point data set using values of 150 (top left) 500 (top right) 1000 (bottom left) and 2500 (bottom right) for the 'radius' of the density function. Grey scale varies. Generated with ArcView spatial analyst.

KDE produces significantly smoother surfaces than the other methods described. Moreover, the kernel can be made asymmetrical so that point distributions that seem to cluster in a directional way can have a more appropriate form of density function. Figure 9.2 shows how changing the form of the kernel—in this case to make the bumps wider—can allow tuning of the resulting density estimate from a highly local surface that reflects smaller clusters of points to a far more general approximator which reflects larger scale structure in the data. Unless great care is taken, however, KDE—like simple density estimates—can be significantly affected by edge effects.

9.3 TREND SURFACE ANALYSIS

One of the simplest and most widely used methods for interpolating data measured at ordinal scale or higher is *trend surface analysis*, which is essentially a polynomial regression technique extended from two to three dimensions. In trend surface analysis, an assumption is made that there is some underlying trend that can be modelled by a polynomial mathematical function, and that the observations are therefore the sum of this and a random error. Trend surface analysis is similar to regression analysis, but extended to 3D (see Figure 9.3). In a simple linear regression, the relationship between a dependent variable Z and independent X would be estimated as:

$$z = a + bx \tag{9.1}$$

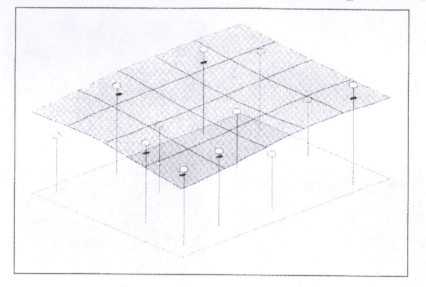

Figure 9.3 Trend surface analysis: the aim is to generate a 'best fit' smooth surface that approaches the data points, and the surface is unlikely to actually pass through any of the points.

where a is the intercept, and b the slope. In trend surface analysis, it is assumed that the spatial variable (z) is the dependent variable, while the co-ordinates x and y are independent variables. In a linear trend surface, the equation relating z to x and y is a surface of the form:

$$z = f(x, y) = a + bx + cy \qquad (9.2)$$

and the coefficients b and c are chosen by the least squares method, which defines the best 'fit' of a surface to the points as that which minimises the sum of the squared deviations from the surface.

As with regression analysis, different orders of polynomial equations can also be used to describe surfaces with increasing complexity. For example, a quadratic equation (an equation that includes terms to the power of 2) describes a simple hill or valley:

$$z = a + bx + cy + dx^2 + exy + fy^2 \qquad (9.3)$$

The highest power of the equation is referred to as the *degree* of the equation, and in general, surfaces that are described by equations of degree n can have at most $n-1$ inflexions. A cubic surface (degree 3) can therefore have one maximum and one minimum:

$$z = a + bx + cy + dx^2 + exy + fy^2 + gx^3 + hx^2 y + ixy^2 + jy^3 \qquad (9.4)$$

Because continuous surfaces cannot be adequately represented in GIS systems, most implementations of trend surface analysis are in raster-based

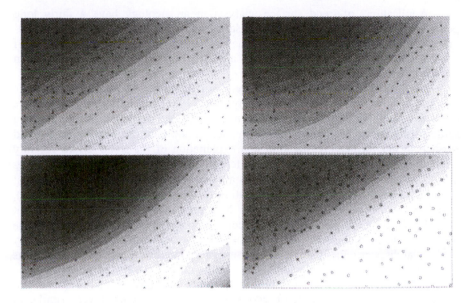

Figure 9.4 Example trend surfaces. Linear (top right), quadratic (top left) and cubic (bottom right) surfaces generated from the same values and a logistic trend surface (bottom left) from the same points, but different attributes, shown as crosses for presence and zeros for absence. Generated with ArcView spatial analyst.

systems, with the output taking the form of a discrete approximation of the continuous surface obtained by calculating the value on the surface of the centre of each grid cell, or the average of the four corners of each grid cell.

One disadvantage of trend surfaces is that the surface described is unconstrained. In order to satisfy the least squares criterion, it can be necessary for intervening values to be interpolated as very high or very low, sometimes orders of magnitude outside the range of the data. Higher order polynomials, in particular, should therefore be used with extreme caution. It should also be noted that values interpolated outside the area defined by the convex hull of the data points—the area bounded by a line connecting the outermost points—should be treated with extreme scepticism, if not discarded as a matter of course. Higher order polynomial trend surfaces can also be very computationally expensive, because their estimation requires the solution of a large number of simultaneous equations.

Given that it is not generally possible or advisable to use higher order polynomials for trend surface analysis, the surfaces described are almost always relatively simple (see Figure 9.4 for examples). As such, trend surface analysis is most meaningfully applied to modelling the spatial distribution of variables that are likely to have relatively simple forms, and it is not appropriate for approximating complex surfaces such as topography.

An advantage of the method is that, like two-dimensional regression analysis, it is possible to analyse the residuals (the distance from observations to surface) to obtain an estimate (r^2) of how well the surface fits the data. It is also useful to map the residuals because spatial patterning within them may reveal systematic problems with the process.

Applications of trend surface analysis in archaeology have a fairly long pedigree. Hodder and Orton (1976) describe the technique and present studies including a trend surface analysis of the distribution of length/width indices of Bagterp spearheads in northern Europe, and of percentages of Oxford pottery in southern Britain.

Another example of the 'global' trend surface approach can be found in the analysis of Lowland Classic Maya sites by Bove (1981) and subsequently Kvamme (1990d). Both identify spatial autocorrelation in the terminal dates of Maya settlement sites (see boxed section in chapter 6) and then use a polynomial trend surface to aid further investigation and explanation of the trend. More recently Neiman (1997) has also investigated the terminal dates of Classic Maya sites using a *loess* model of the same data. Loess is a variation on trend surface analysis that provides a more local and robust estimation of trend—see *e.g.* Cleveland (1993), (cited in Neiman 1997), for discussion. Neiman's interpretation is complex but uses the resulting loess trend surface in comparison to mean annual rainfall for the Maya Lowlands to argue that the collapse phenomenon was caused, ultimately, by ecological disaster rather than drought or invasion. Interestingly, Neiman also turns to variogram methods to understand the structure of the dataset (see below).

Finally, it is possible to fit a probability model (usually a logistic equation) to a set of data presence/absence observations. This is conceptually very similar to 'classic' trend surface analysis but provides output scaled between 0 and 1 and which can be appropriately interpreted as an estimate of the probability that any location will contain one or other of the input categories.

9.4 APPROACHES THAT USE TRIANGULATION

Often simply referred to as 'contouring', because contours can easily be generated by joining together points of equal value on the sides of the triangles, this approach involves generating a triangulation from all the known data values. This is usually done by deriving a Delaunay triangulation, as described in Chapter 5 in the discussion of Triangulated Irregular Network models of elevation. Where locations lie away from data points but within the triangulation—in other words within the *convex hull* of the data points—then their values may be obtained from the geometry of the relevant triangular face (see Figure 9.5).

When unknown values are assumed to lie on flat triangular facets, then the operation is linear, constrained, and behaves in a very predictable manner because unknown values must lie within the range of the three nearest known values (assuming that the interpolation is restricted to the convex hull of the data points). It produces a 'faceted' model that is continuous, although its derivatives are likely to be non-continuous.

If, instead of triangular facets, small polynomial 'patches' are used in order to produce a result with continuous derivatives, then the approach is sometimes referred to as 'non-linear contouring'. This provides a smooth, non-faceted result, and the higher the order of the patch, the higher the order of derivatives that will be continuous. However, it removes the constraint that the unknown values must lie between the three defining data points. Inappropriate choice of parameters or data points can therefore lead to strange and extreme 'folds' in the result. As with higher order polynomial trend surfaces, therefore, the use of non-linear contouring should be treated with caution.

All such approaches are *exact* interpolators because the resulting surface must pass through the data points, and are generally extremely *local*. Unknown values are wholly defined by the closest three known values, and are unaffected by any other data points. In some applications this might be desirable because deviant

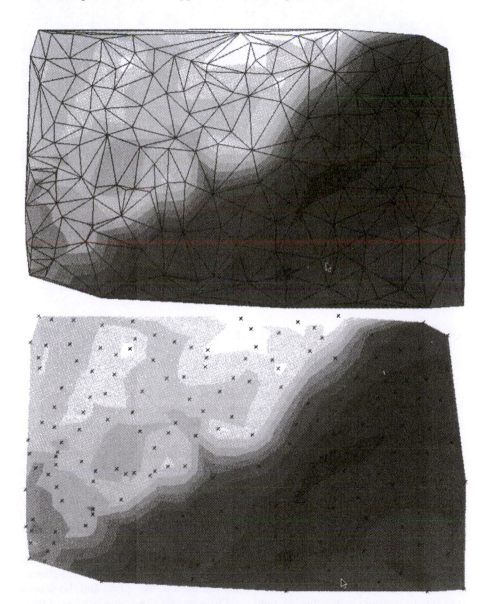

Figure 9.5 Interpolation using Delaunay triangulation. Higher values are shown as lighter colours. The rather angular result is clearer in the bottom image, which also shows the data points from which the surface derives. Note the interpolated values are limited to the convex hull of the data points.

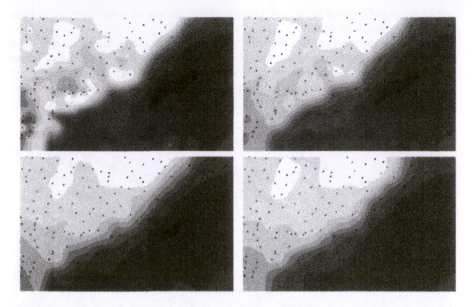

Figure 9.6 Interpolation with splines showing the effect of increasing tension. Values for 'tension' are 0.00025 (top left) 500 (top right) 10,000 (bottom left) and 10,000,000 (bottom right). Generated with ArcView Spatial Analyst.

values will be easily identified but it also means that these approaches can be particularly susceptible to 'noise' in the sample data.

9.5 APPROACHES THAT USE SPLINES

The term *spline* was originally used to describe a flexible ruler used for drawing curves. In contemporary terminology, *spline* curves are interpolators that are related to trend surfaces in that polynomial equations are used to describe a curve. The key difference is that splines are piecewise functions: polynomial functions are fit to small numbers of points which are joined together by constraining the derivatives of the curve where the functions join.

Because the function is, essentially, a local interpolator the polynomials used can be low-order equations, commonly quadratic or cubic. The resulting surface is also controlled by the severity of the constraint placed on the join. A quadratic spline must be constrained so that the first derivative is continuous, a cubic so that at least the second derivative is continuous and so on. More complex forms of splines that 'blend together' individual polynomial equations can also be used, the widely used Bezier- or B-splines being a good example.

Various methods can be used to control the nature of the surface that is fit to the data points. As with numerical averaging (see below) it is common for software to allow the user to choose how many points are used in the spline and the nearest *n* points to the unknown value are then used to generate the spline patches. Lower values of *n* result in a more complex surface while larger values provide a greater degree of smoothing. Additionally, a *tension* value may be applied to the interpolation. Increasing this has an effect similar to increasing the tautness of a rubber sheet, in that the complexity of the surface is reduced and the surface becomes more of an approximator. Lower tension values permit the spline to follow the data points more closely, but very low values can lead to the same kinds of anomalous 'folds' that can result from higher order polynomial equations. Figure 9.6 illustrates the effect of increasing the tension parameter while holding the number of points used constant.

The main advantage of splines is that they can model complex surfaces with relatively low computational expense. This means that they can be quick and visually effective interpolators and, because they are inherently continuous, derivatives such as slope and aspect can be easily calculated. Their main disadvantages are that no estimates of error are given, and that the continuity can mask abrupt changes in the data.

9.6 NUMERICAL APPROXIMATION

An alternative strategy for interpolating values between known locations is to use numerical approximation procedures. These do not constrain the result to pass through the data points, but instead use the points to approximate the values within both sampled and unsampled locations. Using only the *x* dimension (for clarity) a weighted average interpolator can be written as:

$$\hat{Z}(x) = \sum_{i=1}^{n} \lambda_i Z(x_i)$$

$$(9.5)$$

where *n* is the number of points we wish to use to estimate from, and λ_i refers to the weighting given to each point. A requirement of the method is that these weights add up to one, so that:

$$\sum \lambda_i = 1$$

$$(9.6)$$

The points used for the interpolation can be obtained in several ways. Commonly they can be selected either by specifying that the nearest *n* points are used for each unknown value or, alternatively, by specifying a distance *d* within which all points are used. The weightings λ_i are calculated from $\Phi(d(x,x_i))$ where $\Phi(d)$ must become very large as d becomes very small.

The most widely applied approach is that adopted by Shapiro (1993) within the GRASS GIS (r.surf.idw, r.surf.idw2 and s.surf.idw), referred to as *inverse distance weighting*, and is similar to procedures available within a number of other systems. In this interpolator $\Phi(d)$ is the inverse squared distance weighting, so that the interpolator becomes:

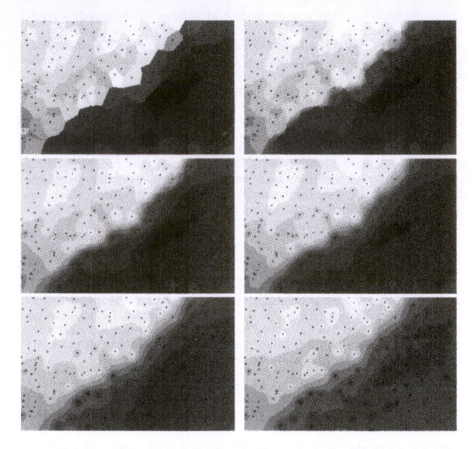

Figure 9.7 Interpolation with Inverse Distance Weighting, showing the effect of increasing *n* (number of points used). Values are 1, 2 (top) 5, 10 (middle) 50 and 176 (bottom). Grey scale is the same for all images in this case. Generated with ArcView Spatial Analyst.

$$\hat{Z}(x_j) = \frac{\sum_{i=1}^{n} \sqrt{d_{ij}} . Z(x_i)}{\sum_{i=1}^{n} \sqrt{d_{ij}}}$$

(9.7)

This works on the principle that unknown values are more likely to resemble near values than distant ones, so the values of each point used to estimate the unknown value are weighted in inverse proportion to their distance from the unknown value. The actual weighting for each data point is the square root of the distance to the unknown point, and these weighted values are then averaged to produce an estimate of the value at the unknown location. The effect is a simple, generally robust approximation procedure for the interpolation of surfaces from a wide variety of point data.

In contrast to the contouring methods described above, interpolation by this method does not rely only on the nearest three data points. Instead it introduces an element of smoothing, the degree of which is controlled by the number of neighbouring points used in the calculation (or the radius *d* within which points are included in the estimate). The larger the number of points used, or the larger the distance, the more of the individual variation within the observations will be smoothed out between the data points. This is, of course, a more general formulation of the inverse distance weighting interpolation we described for interpolation from digitised contours. In that case the points used were defined as eight points found by following cardinal directions to the nearest contour lines.

Figure 9.7 shows the effect of choosing different values of *n* for inverse distance weighting. Note how the surface for a value of 1 is essentially a Voronoi tessellation, with height values attached to the polygons. Surfaces for $n = 2$ upwards become steadily smoother while the very high values of *n* tend to show a very smooth background between the data points, with the points themselves appearing as 'peaks' above this. It is usually necessary to experiment to obtain the best compromise between the 'blocky' result obtained with low values of *n* (or low distances for point inclusion) and the 'peaky' result obtained with very high values.

9.7 GEOSTATISTICS AND KRIGING

The main problem with the use of numerical average methods is that they provide no means for deciding what is the right number of points *n*, or the ideal weights λ_i for any given interpolation. They also provide no indication about how reliable each of the interpolated values is, although it should be clear that the further from the data points the less reliable the interpolated values become.

The term *geostatistics* refers to an approach to spatial variables that addresses these concerns. It was developed by the mathematician George Matheron and D.G. Krige (a South African mining engineer) and it has found wide application in groundwater modelling, soil science, geology, geography and, increasingly, archaeology. The approach is based on the recognition that many spatial variables are best described by a stochastic surface. This is achieved using *regionalized variable theory* (Matheron 1971) that divides the variable into *structure* and *noise*, so that any variable is composed of three components:

1. a structural component described by a function *m*(x);
2. a random component spatially correlated with (1) called $\varepsilon'(x)$;
3. a random noise or residual error not spatially correlated with (1) called ε''.

These are shown diagrammatically in Figure 9.8, which also serves to illustrate the general idea of Kriging, which is to obtain the form of the trend function *m*(x) from the observations. Using these assumptions, and using only one spatial dimension for clarity, the value of a variable *Z* at position *x* can be given by:

$$Z(x) = m(x) + \varepsilon'(x) + \varepsilon''$$

(9.8)

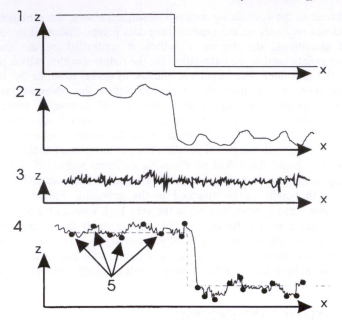

Figure 9.8 Three components of a regionalized variable: (1) structural component, (2) random spatially correlated component and (3) random noise element. (4) Shows how the observations (5) might relate to the other components.

Kriging or *optimal interpolation* is a technique that extends the weighted average method to the optimisation of the values of the weights λ_i to produce the 'Best Linear Unbiased Estimate' (BLUE) of a variable at any given location. It can be used for the same type of data but produces better results in complex spatial variables. The following discussion is heavily based on Burrough (1986:155–165).

Central to the success of geostatistics is the *semivariogram*. This relates the distance between each pair of points (the *lag*) to the influence that these points have upon one another. This is used to determine the underlying structural properties of the variable. In the simplest case of a spatial variable that does not change over the region of interest, then $m(x)$ will be the mean value of the sampling area. In this case the average difference between any two places x and $x+h$ will be zero:

$$E[Z(x) - Z(x+h)] = 0 \tag{9.9}$$

It is assumed that the variance of the differences in Z at the observation points depends only on the spatial distance between the observations so that:

$$E[\{Z(x) - Z(x+h)\}^2] = E[\{\varepsilon'(x) - \varepsilon'(x+h)\}^2] = 2\gamma(h) \tag{9.10}$$

where $\gamma(h)$ is a function known as the *semivariance*—a measure of the extent to which distance h is related to the similarity of any two points. The semivariance can be obtained by estimation from the sample data. If n is the number of *pairs* of points, each separated by distance h, then the estimated value of $\gamma(h)$– written $\hat{\gamma}(h)$– is as follows:

$$\hat{\gamma}(h) = \frac{1}{2n} \sum_{i=1}^{n} \{Z(x_i) - Z(x_i - h)\}^2 \qquad (9.11)$$

This value can then be calculated for all possible lags within the data set, and plotting this estimated value of semivariance against the sample spacing h (often called the 'lag') gives the *semivariogram*.

The main characteristics of the semivariogram are the curve, which often rises at low values of h until it reaches a plateau area referred to as the *sill*. The lag at which the curve stops rising and levels off into the sill is referred to as the *range*. A curve can be fit to the observations by a variety of mathematical models. Commonly spherical, exponential, linear, gaussian equations or combinations of these are used. Models can also be *bounded*, if they reach a sill, or *unbounded* if they do not. The range is important because it represents the maximum lag at which an observation will be likely to influence an interpolated value. Beyond that range, denoted by a, observations cannot be said to influence the interpolated value. An estimate of the uncorrelated noise in the data is obtained from the intercept (the point at which the curve crosses the y-axis). If this *nugget effect* is too great then no model should be fit to the data.

Having obtained the semivariogram for a set of data observations, it is possible to use a technique similar to numerical averaging (see above) to interpolate the unknown values. In the case of *ordinary Kriging*, the weights are obtained from the semivariogram instead of from an arbitrary function. The function to estimate for the value at the unknown location, written $Z(x_0)$, can be written as follows:

$$\hat{Z}(x_0) = \sum_{i=1}^{n} \lambda_i Z(x_i) \qquad (9.12)$$

where (as usual) the known values are numbered from 1 to n. The weight to be ascribed to each known value is given by the λ_i symbol and they must sum to one. The values of λ_i are calculated in such a way as to reflect the distance of the estimator points from the unknown value as observed in the semivariogram. This is done by selecting weights that minimise the estimation variance σ_e^2, which is calculated from:

$$\sigma_e^2 = \sum_{j=1}^{n} \lambda_j \gamma(x_j, x_0) + \Psi \qquad (9.13)$$

where $g(x_j, x_0)$ is the semivariance between the known point and the unknown point (the quantity Y is required for minimalisation). The minimum value of σ_e^2 occurs when:

$$\sum_{j=1}^{n} \lambda_j \gamma\left(x_i, x_j\right) + \Psi = \gamma\left(x_i, x_0\right) \tag{9.14}$$

for all values of i. In this equation $g(x_j,x_i)$ is the semivariance between the sampling points. Both this value and the value $g(x_j,x_0)$ are obtained from the model that was fit to the semivariogram.

There are numerous variants of variograms and Kriging, including methods that investigate or take account of anisotropy in spatial variables or that use other spatial information to assist with interpolation (*stratified Kriging* or *co-Kriging*). These are well beyond the scope of this volume, but further details can be found in Burrough (1986:147-150), Bailey and Gatrell (1995) and Isaaks and Srivastava (1989).

One of the advantages of Kriging methods over simple numerical approximation is that the semivariogram provides an estimate of the *residual noise* in the data. If this residual noise is so large that the structure is not visible then no interpolation technique should be attempted. The semivariogram also provides an optimal estimate of the weighting function if an interpolation is appropriate. A useful by-product of the Kriging equations, the *estimation error* (Kriging variance) can also be mapped and can be a useful indicator of how reliable the interpolated values are at any point. Obviously the estimation error rises with distance from the observations, reflecting the fact that the estimation becomes less reliable.

Mapping the Kriging variance also gives an indication of where the interpolation is most reliable. Figure 9.9 shows both the result of ordinary Kriging, using a linear-plus-sill model and also a map of the Kriging variances. As we might expect, this suggests that the interpolation is most reliable closest to the data points but it also reveals several regions (in darker shades) where we may decide that the interpolated values are likely to be poor estimates.

Against that, Kriging is heavily computational, and in many circumstances the results may not be substantially better than those obtained from using a simpler method. Robinson and Zubrow investigated a range of spatial interpolators for simulated archaeological field data and concluded that:

"Kriging is computationally expensive and, if not fine tuned, it works poorly. Incidentally, it is very expensive in time to calculate and is not recommended unless you have special need and special justification. One justification could be concern about edge effects. Kriging is superior to other techniques in using local topography to infer the form along the edge." (Robinson and Zubrow 1999:79)

It is difficult, however, to take such a generalisation entirely seriously because 'Kriging' covers such a wide range of different specific methods. The example contains no examination of the variogram that should accompany a geostatistical analysis and a more considered conclusion might be that the methods are not amenable to 'button pushing' applications. Herein may actually lie the greatest strength of geostatistics within archaeology—it is not possible to apply the methods without gaining a sound understanding of the structure of the data. Contrary to Zubrow and Robinson's conclusions, other archaeologists have found geostatistical methods, including Kriging, to be useful. Ebert (1998) investigated their use in the analysis of real field data from Als, Denmark, and found two key advantages. Firstly, the methods indeed allow archaeologists to study the *structure* of spatial variation through the variogram, and secondly, they provide the ability to

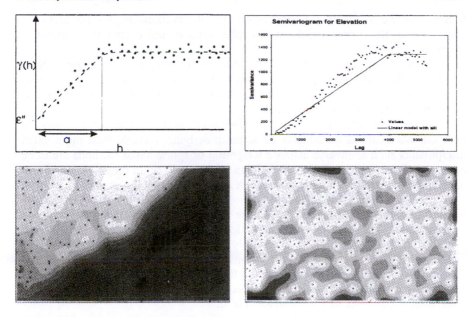

Figure 9.9 The use of the semivariogram in Kriging. Idealised semivariogram (top left) and the actual variogram for the elevation data (top right) with a linear-plus-sill model. Bottom left shows the result of ordinary Kriging using this model and bottom right shows the Kriging variance (higher values are darker). Generated with Marco Boeringa's Kriging Interpolator extension developed for the hydrology department of Amsterdam Water Supply.

produce interpolated maps of spatial distributions and to have confidence in those estimates.

9.8 SUMMARY

This chapter has introduced another large field of spatial technology, commenting on a range of different possible methods for interpolation of continuous variables from point data. It is not possible to 'recommend' one method over another for the analysis of archaeological data because the different characteristics of these methods mean that they are applicable to different archaeological questions. The important thing is to understand the assumptions and nature of the method that is being used, and be certain that it is appropriate for both the data and, more importantly, the archaeological question being posed of it.

We have described several approaches to creating 'density' surfaces from unclassified point observations, although we have also suggested that caution must be exercised whenever continuous or ordinal surfaces are produced from densities. Again, the issue here is that the result must be in some sense meaningful or assist in some way in the interpretation of the distribution. Too often, in published examples, it is not clear that either of these is true.

Trend surface analysis is suitable for the analysis of phenomena that can genuinely be assumed to exhibit simple types of variation over space: 'wave of advance' models, for example, or the quantifiable effects of economic geography.

If it is possible to justify the assumption that observations are the product of a single, relatively simple trend variable plus random noise then the method is suitable.

Where observations are generally free from error, and the closest fit of surface to observations is therefore required, methods based on triangulation will give the most local estimate although linear methods produce rather angular results that generate discontinuous derivatives.

In situations where the modelled surface needs to be complex, and we wish to treat our observations as samples of some underlying variable then more exact interpolators such as inverse distance weighting or spline-based methods are to be preferred. These will give a better estimate of the depth of an underlying layer, or of the changing values of pottery density in a settlement site, for example.

Finally, where a thorough investigation of the spatial structure of the variable is required then archaeologists may choose to invest both time and computation in geostatistics. While Kriging methods undoubtedly provide the best estimates of complex spatial variables, this is certainly at the cost of great complexity. The mathematical and statistical underpinnings of regionalised variable theory and Kriging have been barely touched upon in this chapter, and at the present time, the number of archaeologists who will have the time or inclination to properly understand these methods is very small indeed.

CHAPTER TEN

Visibility analysis and archaeology

"There's more to seeing than meets the eyball." (Chalmers 1999:5)

Ideas of visibility and intervisibility have always been important within archaeological analysis and interpretation. Whether as a factor in the location of 'defended' sites or in the frequently noticed choice of location for prehistoric funerary mounds on 'false crests', the level of visibility of single sites, or intervisibility within a group of sites has long been acknowledged as having played a role in the structuring of archaeological landscapes. What distinguishes recent trends is the gradual move away from little more than a tacit acknowledgement of the importance of visual factors, to instead assigning visibility a central analytical rôle. Visibility, usually equated with perception, has increasingly been viewed as an important correlate to more abstracted approaches to the study of the environment.

Whilst such visual phenomena have long been regarded as important, the acts of 'seeing' and 'looking' have been very hard to operationalise in any traditional, methodological sense. Concepts such as 'panoramic', 'prominent' and 'hidden' may well be useful heuristics, but are very hard to investigate in practice. Visual characteristics have often been mentioned but rarely explored in any formal and meaningful way. As a result, the incorporation of visibility or intervisibility with archaeological interpretations has tended to be anecdotal, at best. Although a number of techniques were developed through the 1970s and 1980s, it can be argued that the first systematic attempts to exploit the visual characteristics, or properties, of locations came in the early 1990s with the widespread adoption by archaeologists of Geographical Information Systems (GIS).

GIS provides archaeologists with a set of standard functions for calculating line-of-sight products from digital models of surface topography. With the discovery of such standard functions for the characterisation of visual parameters nestling within the generic GIS toolbox, researchers found themselves in the unique position of being able to undertake visibility-based studies in a rapid, quantifiable and repeatable way. Through the selection of the appropriate function or command, the precise area of land visible from a given point in a landscape could be identified and potential lines-of-sight between points evaluated. What is more, the results of such analyses could be expressed using the familiar and intuitive medium of the map sheet, so central to archaeological landscape study.

10.1 THE IMPORTANCE OF VISIBILITY IN ARCHAEOLOGICAL ANALYSIS

The visual appearance of a place is, in most cultures and for most people, the most significant impact a location has upon any individual's many senses. A place may feel cold or hot, may smell or have a particular sound quality, but it is the visual characteristics that are most frequently remembered and referred to, and often form the basis upon which a description of a place is instinctively based. It is this role in constructing a sense of place that goes a long way towards explaining the recent resurgence of interest in visibility in archaeological analysis. In some instances, visibility can be regarded as a key factor in attempting to answer the question as to why a particular site is in a particular place, rather than all the other places it might have been located.

Of course, visibility is not the only factor at work in the selection of locations for settlements, monuments or other activities. An almost infinite number of factors may have been considerations. Locations may be chosen because they are close to particular practical or symbolic resources or to existing sites. They may be selected for less systematic reasons: for example being gradually defined and individualised through repeated experience and encounter, or as a result of their role as anchoring or nodal points in complex webs of myth and folk-tale (Basso 1996).

The visual characteristics of a site (whatever its significance or presumed function) can be highly localised, such as within caves or buildings where the visual appearance is highly controlled, compelling the visitor to contemplate only the site in question. Alternatively the visual characteristics of a site can derive from its position within a wider visual landscape in which it may have relationships of visibility with other contemporary (or earlier) sites, or with natural components of either the regional landscape or the wider universe. Such relationships may be deliberate, and therefore meaningful to those who chose to establish a site. They can also be accidental and, as a result, not meaningful, although subject to gaining meaning through time as stories and explanations are offered as part of the construction of a wider cosmology. Visibility can be important between related or similar sites, as in the case of watchtowers or signalling systems, or between different types of site as seems to be the case where megaliths have been aligned on astronomical features or events.

10.2 ARCHAEOLOGICAL APPROACHES TO VISIBILITY

With the notable exception of astronomical alignments, the analysis of visibility almost entirely escaped the quantitative revolution of the New Archaeology. As discussed in Chapter 1, during the 1970s archaeology adopted formal spatial analysis wholesale (Hodder and Orton 1976, Clarke 1977a). In saying this it is important to emphasise that this adoption tended to concentrate on elaborating the statistical relationships between cultural features or finds (for example the distribution of artefacts with respect to a given production site), or on relationships between cultural features and environment, such as settlements and raw materials. Despite the acknowledged importance of visibility within archaeological interpretation, and with one or two exceptions, the New Archaeology rarely had the will or the tools to develop formal methods for visibility analysis.

One notable exception to the New Archaeology's apparent lack of interest in visibility analysis can be seen in studies of the prehistoric monuments of Orkney (Renfrew 1979, Fraser 1983). Here, visibility was posited as a significant factor in the location of prehistoric stone cairns that were supposedly sited so as to delineate discrete territories or activity areas in the wider landscape. The desire to seek 'territories' in this landscape reflected similar interpretations of Neolithic and Bronze Age Wessex that saw Neolithic long barrows as markers for family territories which were themselves the early stages of development towards increasingly hierarchical social systems.

The methodology for these studies involved repeated observations from fixed points in the landscape that were collated to produce maps with overlapping zones of decreasing/increasing visibility. The resulting patterns of intervisibility between cairns were then used as quantified variables (along with other factors such as slope, drainage, rockiness and distance from the coast) to explain the observed distribution of cairn locations (Renfrew 1979:13-20).

This approach was further refined by Fraser, in one of the most thorough archaeological treatments of the issue of visibility (Fraser 1983). This introduced the notion of visibility classes, in an attempt to standardise the characterisation of fields-of-view. A series of visibility-classes were identified, defined as 'Distant' (visibility exceeds 5km), 'Intermediate' (between 500m and 5km), and 'Restricted' (less than 500m). This choice of classes was largely contingent, and Fraser was careful to state that the methodology was designed to be reproducible at the localised scale, partitioning the visible horizon into distinctive sections that would be consistently recognisable throughout Orkney (Fraser 1983:298-303). Despite this proviso, it is interesting to note that there has been a recent resurgence of interest in his methodology in the context of rock-art studies, particularly relating to the tendency for certain petroglyphs to be located at points offering wide views of the surrounding countryside (Bradley 1991:137-138, Bradley 1995).

Several significant factors can be identified in these studies. In making a formal attempt to incorporate vision into analyses, the perceptual and essentially human acts of seeing and looking were translated into the heavily simplified concepts of field-of-view and line-of-sight. These were always recorded with respect to locations *fixed* in the landscape, were quantifiable and, as a result of the introduction of concepts such as visibility-class, the product of a rigorous and reproducible methodology. Once defined, the quantified view-character of each site was utilised as one of a number of environmental criteria for the examination of the location and observed patterning of structures within the landscape. Although the specific interpretations given to such results are debatable, the work of Renfrew and Fraser is notable in that it used notions of a 'field-of-view' (the area visible from a fixed location) and 'intervisibility' (presence or absence of a line-of-sight between features). As we will see, these are precisely the building blocks that have been used more recently within GIS-based studies.

A second strand in debates concerning the role of vision in archaeological research emerged as a result of developments in archaeological theory in the late 1980s and early 1990s. Central to these theoretical debates was a growing appreciation of the role of past peoples as active, purposeful agents situated within a meaningful world. A resurgence of interest in the body was coupled with an emphasis upon partial, situated and fragmentary knowledges as opposed to any holistic understandings of the inhabited world. Perception was seen as bodily centred and culturally embedded rather than an abstracted consequence of the environment and appreciation was given over to experience on the human scale (Tilley 1993:56).

Such approaches have paid particular attention to the study of prehistoric monumentality. Previously analysed in terms of typological or spatial grouping, many of the new approaches stressed an understanding of monuments as physical components of a lived environment, where the ordering of space and access, both physical and visual, were seen as of critical importance. Tilley, for example, advocated a more embodied approach, emphasising the importance of bodily movement within and through monuments and landscapes (Tilley 1994:73). His analysis of the Long Barrows of Cranbourne Chase (ibid:156-166) was undertaken by visiting the sites, observing and moving around them, and interpolating lines of sight from a number of points within the landscape. Approaches such as this, informed by phenomenology, move the analytical emphasis away from fixed-point fields-of-view and point-to-point intervisibility. Instead, vision is considered as the act of a mobile body and 'viewsheds' are replaced by sequences of visual cues and moments of panoramic display or obscurement, linked to bodily engagement and movement (Devereux 1991, Thomas 1993:37, Barrett 1994, Tilley 1994).

In practical terms such studies are notable for the lack of any formal methodology. In moving away from simplification and abstraction in favour of embracing the full and highly nuanced perceptual complexity of vision, there has been a return to the more descriptive and anecdotal methods of recording such effects typical of the early antiquarians. Accounts rely upon thick description, often in the form of a travelogue, punctuated with photographs and frequent perspectival drawings. It should be noted that within such schema, factors such as field-of-view and intervisibility are still seen as useful heuristics. For example, in a seminal discussion regarding place, space and perceptions of the landscape, Evans called for the creation of an index of the visibility and appearance of features in the landscape. He proposed the concept of 'visual ranges' framed by topographic features, which mirrors closely the pre-existing notion of a field-of-view suggested by Renfrew. He also laid emphasis upon how sites and monuments interacted as a visual network that was seen as a potential factor in their historic generation (Evans 1985). In a recent discussion as to the visual-spatial relationship between megaliths and prominent natural formations such as rock-outcrops, patterns of intervisibility were examined in relation to the way in which highly prominent natural features served to key megaliths into the landscape and emphasise their character as special places (Tilley 1994:99). Despite the fact that such factors are seen as important, the formal techniques proposed by researchers such as Renfrew and Fraser did not find frequent application within this body of work.

10.3 HOW DOES THE GIS CALCULATE VISIBILITY?

The GIS offers two techniques for incorporating visibility into an archaeological analysis. These take the form of viewshed calculation, which determines what areas of the study area can theoretically be seen from a given viewing location, and the determination of direct intervisibility between a set of features.

The calculation of a line of sight map or viewshed for a location is a relatively trivial raster-based computing problem and is available within the current functionality of many GIS (for example Arc Info's *Visibility* command, the *Calculate Viewshed* command within the ArcView Spatial Analyst and the *Analysis/Context operators/Viewshed* command in Idrisi). To calculate a viewshed you need the following thematic layers of information: a DEM and a layer encoding the location you wish to determine the viewshed from. In the case of systems such as Arc Info and Arcview this can correspond to either a discrete point or linear feature, whilst systems such as Idrisi and GRASS also include area features.

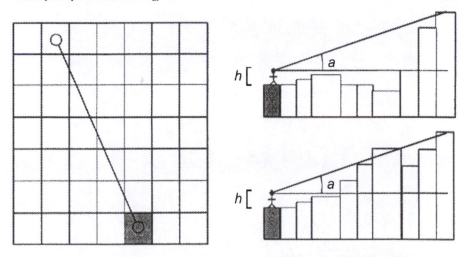

Figure 10.1 Testing for the intervisibility of two cells in an altitude matrix, with observer of height *h*. Some algorithms return a binary result (1 for visible, 0 for not). Others return the angle *a* for visible grid cells.

The actual calculation requires that, for each cell in the raster, a straight line be interpolated between the source point and every other cell within the elevation model. The heights of all the cells which occur on the straight line between the source and target cells can then be obtained in order to ascertain whether or not the cell exceeds the height of the three dimensional line at that point. Figure 10.1 shows diagrammatically how this operates: the example top right shows the case for the existence of a line of sight, while the example bottom right shows the case for no line of sight.

The result of each of these discrete calculations is either a positive (Figure 10.1, top right) or negative (Figure 10.1, bottom right) result, conventionally coded as a 1 for a visible cell or a 0 for a cell that is not visible. When performed for the entire raster, the result is a binary image with those areas of the landscape that have a direct line of sight from the target cell coded as a 1 and those with no line of sight with 0. Such images are commonly referred to as *visibility maps* or *viewshed maps*.

A typical example is given in Figure 10.2. Many systems will allow restrictions to be imposed on the grid-cells to be included in the calculation such as limiting the distance that grid-cells will be included or the angle of view. Most GIS also allow specification of a notional height above the surface of the DEM for the viewer. In the case of an adult human being a standard offset height of 1.7 metres is often used, although the height of a human eye above the ground does rather depend on the particular human. If we wish to investigate the viewshed for a structure such as a watchtower, then the height of the viewing platform can be used instead.

Some systems also return the *angle* of the line of sight for grid cells that are visible, from which the binary viewshed can be obtained by reclassification if needed and this can be archaeologically interesting in situations where we wish to investigate the visual prominence of one or more monuments. Here, it may be significant whether a viewer looks up or down at a site or monument. A site that is

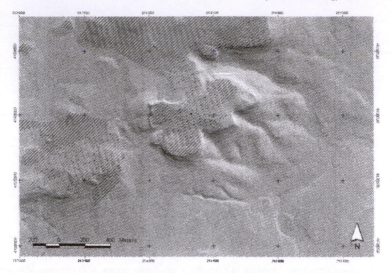

Figure 10.2 The binary 'viewshed' (shaded) of an observer (the dot) inside the settlement of El Gandul, Andalucia. Shown draped over the hillshading map. Note how topography restricts the view in all directions except north and west.

above the viewer's line of sight may have more visual impact, being more likely to be framed against the sky, or it may be symbolically significant. An example is given in Figure 10.3, where the angle of view has been used to reclassify the viewsheds of four neolithic long barrows according to whether they are viewed above, broadly level with or below the viewer's line of sight. Clearly, in this case, the visual impact of the monuments is likely to be greatest in those areas from which the barrows are 'looked up to' (shown in a lighter shade of grey).

Viewshed maps have found frequent application within archaeology. An early example is the work of Gaffney and Stancic, who in their investigations of the Greek period of the Dalmatian island of Hvar, generated a viewshed map for the watchtower at the site of Maslonovik. This then demonstrated that a similar watchtower at Tor would have been visible from Maslinovik, which would in turn have been able to pass warnings to the town of Pharos. Their result supported the assumption that such towers formed "an integral system connected to the town and Pharos whereby watch was kept for any approaching danger" (Gaffney and Stancic, 1991:78).

10.4 VISIBILITY WITHIN SAMPLES OF SITES—THE CUMULATIVE VIEWSHED

In situations where patterns of visibility within a group of sites is of interest, it is possible to obtain a viewshed map for each site location. These individual maps can then be summed, using simple map algebra techniques, to create one surface. This resultant surface represents for each cell within the landscape, the *number* of sites with a line of sight from that cell. For a sample of n sites, the value of this

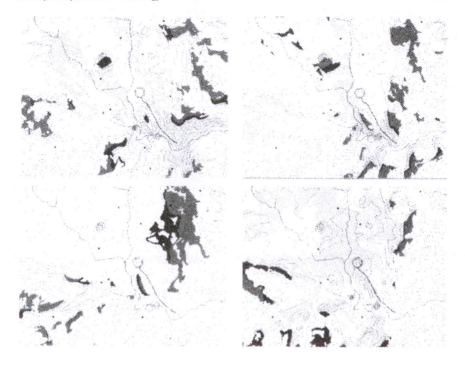

Figure 10.3 Viewshed maps of the Avebury region, southern England, showing the visibility of four long mounds according to whether they are above, broadly level with, or below the viewer's position (Clockwise from top left: East Kennet, West Kennet, Winterbourne Monkton, Beckhampton Road).

surface will obviously consist of integers which are constrained to vary between 0 and *n*. Such a map may be referred to as a *cumulative viewshed map* for the sample of sites within the particular region of interest. The phrase *cumulative viewshed map* is used here to distinguish this *summed* result from the *multiple viewshed map* referred to by Ruggles *et al.* (1993) which is the logical union of a series of individual viewshed maps.

10.5 VISIBILITY OF GROUPS OF SITES—MULTIPLE AND CUMULATIVE VIEWSHEDS

In situations where patterns of visibility within a group of sites is of interest, it is obviously possible to generate a viewshed map for each site location, but this can result in very large numbers of viewshed themes, and their relationships can be difficult to interpret. As a result, it is useful to be able to generate summary themes of the visisbility characteristics of entire groups of sites.

The union of a series of viewshed themes (in which a value 1 represents cells visible from any of the source cells while zero represents locations that are not visible from any) can be termed a *multiple viewshed* theme (Ruggles *et al.* 1993, Ruggles and Medyckyj-Scott 1996). Individual viewshed themes can also be

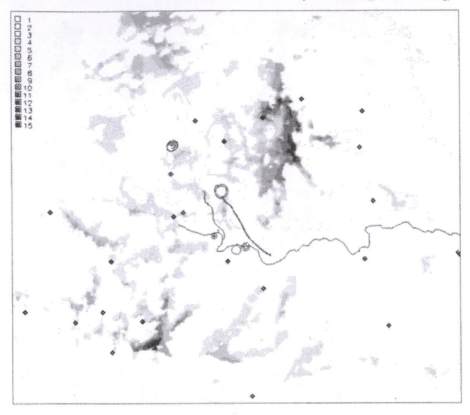

Figure 10.4 Cumulative viewshed map generated from locations of 27 earlier neolithic long mounds and barrows (diamonds) of the Avebury region, in relation to some of the later neolithic monuments and the river Kennet (Wheatley 1996b). Scale represents the number of visible sites.

summed, using map algebra, to create a *cumulative viewshed* theme[6]. This represents, in a raster, the *number* of grid cells with a line of sight from the source location (see Figure 10.4). For a sample of n sites, the value of this will obviously consist of integers constrained to vary between 0 and n. Both of these results can provide useful visual summaries of the way in which a group of sites can relate to other elements of the landscape.

These themes can also be used for more formal approaches to spatial analysis, such as spatial tests of association that we discuss in Chapter 6, and allow systematic approaches to questions relating to the distribution of one group of sites with respect to the visibility of another. It is also possible to perform a point-select operation on this surface based on the locations of the sites from which the cumulative or multiple viewshed was generated. Doing this generates, for each site,

---

[6] Routines do exist to calculate cumulative and multiple viewsheds without the need to calculate the individual files. One example is the *r.cva* module for GRASS GIS, written by Mark Lake. This adds several viewshed improvements to GRASS, and repairs a fault in the r.los routine whereby the viewer height is rounded to an integer (Lake *et al.* 1998), also http://www.ucl.ac.uk/~tcrnmar/ (accessed 3/4/2001).

the number of other sites visible from it, permitting tests of intervisibility to be formulated. An example of this is the analysis of long mounds of the Wessex chalklands, southern England (Wheatley 1995a, 1996b), part of which we discussed in a separate example.

10.6 PROBLEMS WITH VIEWSHED ANALYSIS

There are a number of methodological problems that should be considered when using this type of analysis, some of which may result in errors. We will be discussing one of these, the underlying assumption of reciprocity, in detail in the next section. In undertaking a cumulative viewshed analysis it should always be remembered that rejection of the null hypothesis in the test shows *association* between the monuments and areas of high visibility and does not show that one *caused* the other. For example, it may be that a particular group of sites are situated on high ground because this is dryer. This will be observable by testing the association between topographic height (or distance to rivers) and the locations of the sites. However, it is also the case that high locations tend to have greater visibility, and therefore statistically are more likely to have other sites visible from them—a cumulative viewshed analysis may therefore (correctly) show an *association* between the monuments and areas of high intervisibility even though this had no influence on the monument constructors.

This is, more generally, a problem of *spurious correlation* and it is not unique to the formal analysis of visibility. It is actually a common problem that arises when two things (in this case monument location and intervisibility) are actually both related to a third factor (in this case height) that is the causative part of the relationship (Shennan 1997:121-125).

Another potential problem with cumulative viewshed analysis is that it is very prone to *edge effects*. Where a monument distribution continues outside of the study area, then the cumulative viewshed map may considerably underestimate the number of lines of sites towards the edge of the study region because it does not consider those monuments just outside of the boundary. As a result, it is necessary to be very careful with the choice of study area, perhaps ensuring that the group of monuments is a genuine spatial cluster (see Chapter 6), or generating the cumulative viewshed for a larger area than is used for the statistical test to minimise edge distortions.

Finally, it should be apparent that the main determinant of the viewshed calculation is the digital elevation model. As such, the accuracy and precision of the digital elevation model are absolutely vital to the formal analysis of vision. This issue is further complicated because the overall accuracy of the elevation model (often expressed as *rms* error) may not have much effect on the visibility result, rather a small subset of the elevation values—particularly those on hillcrests—that most impact on the calculated viewshed results. Moreover, as Figure 10.5 clearly shows, the location of the errors with respect to the source location for the calculation will have a significant effect on how much they impact on the results.

Acknowledging the variety of errors that inevitably exist in the terrain model, Fisher (1991) has argued that viewsheds are best considered not as a simple Boolean *in-view*: *out-of-view* layer, but as a probability surface, in which cell

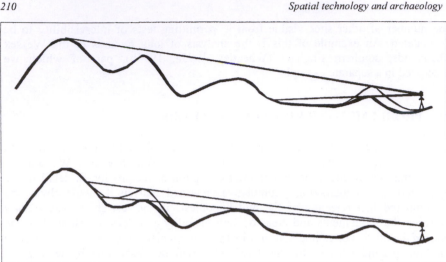

Figure 10.5 Variations in elevation close to the observer (top) clearly have the potential to have a far greater effect on the calculated viewshed than variation farther away (bottom).

values should represent the *probability* that a given target is visible from the source. This has been implemented using monte-carlo simulation techniques in the form of the 'probable viewshed'. In this process, rather than generating a single binary viewshed, a number of such viewsheds are generated for each source location, with random errors added to the elevation model prior to each calculation. These are then summed to generate the final viewshed. In addition to the probable viewshed, Fisher has undertaken important work in the examination of the fall-off of visual clarity with distance, through the generation of 'fuzzy' viewsheds, and the identification of unique terrain phenomenon, such as horizons, within viewsheds (Fisher 1992, 1994, 1995, 1996a, 1996b).

10.7 INTERVISIBILITY AND RECIPROCITY

When a viewshed purports to show all of the areas of the landscape that are visible from a given viewing location, it is important to acknowledge that the converse is not necessarily true in any situation in which a viewer height is specified, and is non-negligible. This issue of reciprocity of view calculations is important, as in many archaeological applications patterns of intervisibility tend to be derived as secondary products from a given viewshed. Cumulative viewshed analysis is a good case in point.

Let us develop this point in the case of a viewshed which is generated to explore line-of-sight between two suspected watchtowers, marked A and B. They could equally correspond to funerary mounds or any other type of site or feature. Imagine the situation shown in Figure 10.6. We could begin by generating a viewshed from location A, specifying a combined watchtower and observer height of 8 metres to reflect the elevated position of our notional viewer. As the figure shows, the location of watchtower B is out of the viewshed, masked by an area of intervening high ground.

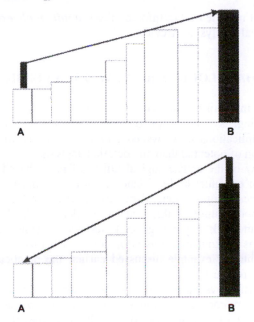

Figure 10.6 The issue of reciprocity in viewshed calculations: where a viewer height is non-negligible, we cannot assume that visibility of B from A implies visibility of A from B.

Despite the fact that we can clearly trace a line of sight between the tops of the two towers, if we take the result of the GIS analysis at face value then we would be forced to conclude that the towers were not in fact intervisible. What we are failing to acknowledge is that the object being viewed is also elevated above the surface of the DEM and if we are to derive conclusions regarding intervisibility from the calculated viewshed, then this has to be taken into consideration when undertaking the original calculation.

The watchtower example is perhaps obvious, but as was noted archaeologically as early as 1983 by Fraser in his pioneering work on Orkney (Fraser 1983), and in the GIS context by Fisher (1996b:1298), the same is true when we are dealing with human observers. What is clear is that a viewshed generated to identify the area of land visible from a given point may be quantitatively different from a reciprocal viewshed generated to identify from where in the landscape that point would be visible. In the first example we need only specify an offset (*e.g.* 1.7m) for the height of the observer at the viewpoint, in the second we need to specify no offset for the viewpoint but instead an offset of 1.7 metres for *all* of the other cells in the DEM. In a useful discussion Loots has stressed this differentiation through the use of two distinct terms to describe each class of viewshed: projective 'views-from' and reflective 'views-to' (Loots 1997).

It is worth noting that whilst some GIS, such as Arc Info and GRASS, provide considerable user control over specifying viewpoint and target cell offsets for viewshed calculations, others do not. In addition, some GIS also provide dedicated tools for determining intervisibility upon a given DEM or TIN without the need to generate a viewshed. A good example is the *Surfacesighting* command within the Arcplot module of Arc Info and the *visibility tools extension* which can be run under Spatial Analyst in Arcview.

10.8 HOW ARCHAEOLOGISTS HAVE APPLIED VISIBILITY ANALYSES

Having discussed in detail how the GIS calculates visibility it is important to look at how archaeologists have applied these techniques. The first wave of archaeological applications of viewshed and intervisibility functions comprised more an illustration of potential than any detailed analysis.

In his review of the archaeological utility of raster-based GIS, van Leusen used viewsheds to examine the local landscapes of a number of Palaeolithic and Mesolithic sites located in the Mergelland Oost area of the Netherlands. The clear 'potential-of-method' nature of this work is evident in the lack of any formal interpretative framework beyond general statements concerning the possible function of sites as lookout locations (van Leusen 1993:4). As with the Gaffney and Stancic watchtower example discussed earlier, the practical application was simple and straightforward.

Little discussion was given over to possible shortcomings in either the technical procedure or conceptual approach, though this is perhaps inevitable given the pioneering nature of such studies. At the other extreme, and in direct contrast, is the work of Ruggles *et al.*, exploring the topographical and astronomical significance of prehistoric stone rows. Here, effort was focused exclusively upon the archaeological and procedural considerations *underlying* any attempt to employ GIS-based visibility approaches in a meaningful way (Ruggles *et al.* 1993). This emphasis is perhaps not surprising given that the study was embedded within the sub-discipline of archaeoastronomy, a field with a pre-existing series of specialised methodologies for the examination of visual phenomenon (Ruggles *et al.* 1991).

These two emerging strands were unified in the major wave of visibility-based analyses which appeared in the literature in 1995 and 1996. Although simple functionalist applications persisted, the majority of these studies can be characterised by a concern to embed visibility more firmly within the archaeological problematic and to devote much more consideration to the inherent limitations of the approach. Such studies are also distinctive for the bold links that are made between developing theoretical issues and the visibility functions available in the generic GIS toolbox.

Typical of these more recent applications are the work of Lock and Harris (1996), and Wheatley (1996b) on the spatial-visual distribution of Prehistoric funerary monuments. In a detailed study of the Danebury region of Southern England, Lock and Harris utilised GIS-based techniques to re-examine a set of interpretations advocated by researchers such as Renfrew in the early 1970s, which emphasised the role of such structures as territorial markers. Based upon the assumption that to act as such a marker a monument had to be visible, an attempt was made to utilise viewshed calculations to delineate supposed territories. These were then directly compared with territories derived from more traditional spatial-analytical techniques such as Thiessen polygon analysis, that made no allowance for visual characteristics (Lock and Harris 1996:223-225).

VISIBILITY FROM A SINGLE SITE: THE PEEL GAP TURRET

Visibility analysis was used to investigate an enigmatic feature located upon the line of the Roman military frontier of Hadrian's Wall, located in the North of England. The structure at Peel Gap comprises a small, rectangular building added to the back face of Hadrian's Wall shortly after its construction in the early 2nd century AD. For a simple account of the feature and the broader context of Hadrian's Wall see Johnson (1989:61-62). In its size and ground plan the structure resembles closely the regular series of turrets that characterise the whole length of the wall. These have traditionally been interpreted as observation posts, located to maximise the views north beyond the frontier, and to either side along the line of the barrier.

The location of the Peel Gap tower appears to contradict this notion of optimum defensive location, as it is at the base of a pass through the dramatic crags that characterise the central sector of the wall zone, rather than on the higher ground to be found at either side.

The first stage in the process was to create a DEM of the Peel Gap study area. This took the form of a 1m resolution DEM based upon a detailed micro-topographic survey of the immediate surrounding landscape. In addition, a vector point layer was created representing the location of the suspected turret, and a visibility analysis was undertaken, using a viewpoint offset of 11 metres above present ground level to represent the combined original height of the turret and observer.

The results of the visibility analysis are shown in Figure 10.7 (right) and confirm strongly the reservations expressed by authors such as Johnson (1989:62) as to the questionable strategic significance of the tower in its present position. The visible area (shaded dark grey) emphasises the heavily restricted view to the north of the Wall towards Scotland and along its line, with the views yielded by the tower location instead appearing to focus upon the natural pass itself, and the area to the rear of the wall zone.

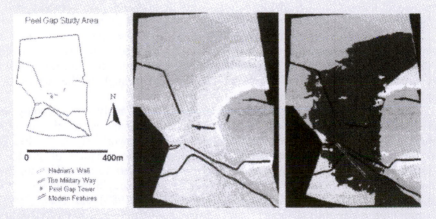

Figure 10.7 The Peel Gap study area (left); 1m resolution DEM (darker grey indicates higher ground) (right); and the viewshed generated with an observer offset of 11 metres. Generated with Arc Info.Topographic data was collected by Richard Bayliss of the University of Newcastle upon Tyne.

As well as work concerned with the visual characteristics of point locations, a number of studies have begun to explore emerging concerns within archaeological theory. Here, an explicit attempt has been made to use GIS-based techniques to explore vision as a deliberate perceptual act, rather than environmental side-effect. In his study of hunting and gathering communities in Southeast Alaska, Maschner has incorporated viewsheds as a cognitive variable in the examination of settlement location (Maschner 1996b), whilst in a study of the pre-Roman town of Nepi in South Etruria, viewsheds have been used to explore the liminality of tomb locations with respect to the town — spatially distant and inaccessible, yet visually immediate (Belcher *et al.* 1999).

One interesting application is that of Smith, who has used viewshed analysis to explore the 2nd century author Pausanias' conceptual map of the city of Piraeus (Smith 1995:243-245).

One of the most detailed explorations of the relationship between cognition, perception and GIS-based visibility studies is the work of Gaffney *et al.* on rock art sites in southwestern Scotland (Gaffney *et al.* 1995). In this study, cumulative viewsheds were used to identify zones in the landscape of increased monument awareness, with the density of such information seen as a direct correlate of socio-symbolic importance. A detailed study of the intervisibility of individual classes of monument locations was then undertaken with respect to this cumulative map. Central to the analysis was the claim that such an approach yields "a mapable, spatially variable index of perception" (ibid:222), study of which would lead to an understanding of the cognitive landscape within which such monuments were components.

10.9 CRITIQUES AND DEVELOPMENTS

GIS-based visibility analyses offer archaeologists a valuable set of analytical tools, but to maximise the potential of these tools they need to be applied within a critical framework, tailored to the unique requirements of archaeology. In recent years archaeological GIS practitioners have begun to address both of these factors.

In a recent review (Wheatley and Gillings 2000) we have highlighted a number of key critiques and issues that need to be addressed or at least considered in any GIS viewshed study. These are summarised in Table 10.1 and range from the impact of vegetation and effects of DEM errors through to the theoretical implications of attempting to *map* aspects of perception.

Following the identification of cumulative viewshed analysis, archaeologists have also continued to tailor GIS-based visibility analyses to the unique requirements of the discipline. Good examples are the work of Loots already mentioned and the pioneering work of Llobera in the creation of gradient viewsheds, which encode degree of obstruction and barriers to visibility into the viewshed (Llobera 1996). A number of authors have also sought to build upon the visual ranges proposed by archaeologists such as Fraser (1983) and Evans (1985) incorporating issues such as object : background clarity and the decay of visibility with increasing distance (see, for example, the 'Higuchi' viewshed proposed by Wheatley and Gillings 2000).

INTERVISIBILITY WITHIN A GROUP OF SITES: LONG MOUNDS OF SALISBURY PLAIN

The earthern long mounds are among the earliest earthworks built in southern England dating to the earlier part of the Neolithic (around 4000BC to about 2500BC) and discussions of their locations have frequently raised issues of monument visibility and intervisibility. To explore their visibility more formally, the locations of thirty-one mounds in a 20km by 20km area around Stonehenge were imported for analysis in the Idrisi GIS. Elevation models were interpolated from digital height data and represented as a 250x250 raster, giving a cell size of 80m (Wheatley 1995a).

Individual viewshed maps were generated for each mound and the areas 'in view' were compared with view areas for random groups of points drawn from the same region. This suggested that the mounds have higher than expected viewshed areas. Secondly, the individual viewsheds were summed to obtain a *cumulative viewshed* for the group of mounds. The number of *other* mounds visible from each mound was then obtained (using attribute extraction) and these values were tabulated, along with the values for the cumulative viewshed itself so that a one-sample Kolmogorov-Smirnov test could be conducted. The mound locations were not found to be random with respect to the overall visibility of other mounds (significant at the 0.05 significance level. See Chapter 6 for details of significance tests).

In simple terms, this suggests that the mounds tend to occur in locations which are visibly prominent, and from which other barrows are visible—seen in the cumulative percentage graph in Figure 10.8 (right). Of course, this does not necessarily mean that this is the reason for the location of the mounds: the association could be to some other factor, such as terrain height, or ridges, which are themselves associated with high visibility.

However, other than unsubstantiated observations, this does provide the first evidence that visibility or intervisibility or both may have been a consideration in the selection of locations for the construction of these early monuments. Archaeologically, this has implications for the interpretation of these mounds as territorial markers, or as 'family' tombs: perhaps suggesting that they are better interpreted as foci for ritual authority, deliberately 'appropriating' pre-existing foci through the incorporation of visual references to them.

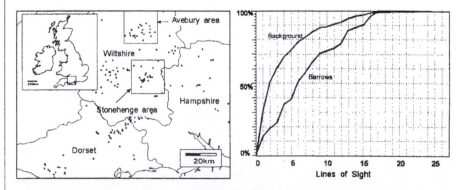

Figure 10.8 The area analysed (left) and the cumulative percentage graph (right) used in the study. The graph suggests that the mounds tend to be in locations from which a higher than expected number of other mounds can be seen.

Table 10.1 Critical issues in GIS-based visibility analyses (Wheatley and Gillings 2000).

Pragmatic Issues:
Palaeoenvironment/palaeovegetation
Object-Background clarity
Mobility
Temporal and Cyclical dynamics
View reciprocity
Simplification

Procedural Issues:
The digital elevation model (DEM)
Algorithms used to calculate line of site and viewshed
The undifferentiated nature of the viewshed
Quantitative rigour
Robustness and sensitivity
Edge effects

Theoretical Issues:
Meta-critique of technological determinism
Occularism, visualism and the nature of perception
The map perspective and the viewing platform

CHAPTER ELEVEN
Cultural resource management

"Naturally, the expansion of GIS usage in UK archaeology will be influenced as much by the nature of the archaeological record and the organizational structure of archaeology in this country, as by the many individual decisions concerning whether to adopt or to reject these ideas." (Harris and Lock 1990)

Since the early 1990s, spatial technologies, specifically Geographic Information Systems, have been widely touted as the ideal tool for the management of heritage information. They were adopted initially in the United States, mainly for their ability to generate predictive models (see Chapter 8), and subsequently in the United Kingdom (*e.g.* Harris 1986), Europe (*e.g.* Roorder and Wiemer 1992) and other parts of the world. Since then, an ever increasing number of archaeological management organisations have become convinced that GIS is the natural successor, or partner, to existing database management systems.

11.1 THE IMPORTANCE OF SPATIAL TECHNOLOGY FOR HERITAGE MANAGEMENT

To understand why this is, we need to recall some of the advantages of storing spatial information in a structured way. We discussed the technical side of this at some length in Chapter 2, but it is helpful to reflect on the implications of this for archaeological information. Most importantly, we should remember that archaeological records themselves have an inherent spatial component. Archaeological sites are always somewhere—although, admittedly, we may not always know where—and they are often poorly represented unless we can record their spatial extents.

Conventional databases allow recording of the attribute component of the heritage record, but they are fundamentally limited with respect to the recording of spatial and topological components. This is particularly problematic where, for example, the record holds information about sites or monuments, which cannot be effectively recorded by a single grid reference. *Linear monuments*, for example—which include earthworks, field systems, roads and many other cultural features—present particular problems to conventional database systems. While these can be represented in a conventional database by including lots of records at intervals, there is no method of indicating, for example, how they join or cross. Many archaeological records choose to artificially divide problem monuments like these into smaller units, and give each unit a different 'site' identifier, but this is messy and can lead to very misleading responses to the simplest kinds of queries ('how many sites are there …').

A second problem is how to store information about archaeologically sensitive areas. In the UK, designations that apply to areas rather than specific sites include Sites of Special Scientific Interest and Conservation Areas, while in the US these would include National Parks. Beyond national designations, there are an increasing number of World Heritage Sites, whose presence may not have statutory implications, but whose inclusion within national or regional heritage records is clearly desirable. Areas present the problem that even if we *could* represent them effectively in a conventional database, the kinds of queries that archaeologists are likely to want answers to (*e.g.* 'how many extant tombs lie within 1km of the boundary of a National Park?') are extremely difficult to express without an explicit coding of the spatial and topological component of the data.

Retrieval and display

The first reason why spatial technologies have proved so popular for the management of archaeological information is found in the kinds of queries that archaeologists wish to make of their data. Arguably the majority are either wholly, or partly, spatial in nature, and consequently are far easier to express with the interface capabilities of a GIS. Retrieval on 'pure' spatial criteria is far easier in GIS than a conventional database—we may wish, for example, to extract a list of all known monuments within 500m of a proposed road development. In a conventional system this could take hours of calculating co-ordinates and formulating queries, while using graphically-oriented GIS software this can be as simple as digitising the road plan, generating a buffer and performing a single map algebra operation.

At the same time, the ability to visualise the location and extent of archaeological materials is a liberating experience for many data managers (although see comments below on data quality) and the identification of erroneously recorded data can often be considerably speeded up using the interactive display capabilities of many GIS. This is largely because the iterative process of:

1.   construct a query
2.   run it
3.   find that it gives the wrong result and
4.   go to (1)

by which most archaeological managers operate can be acknowledged, implemented and legitimised within the GIS as 'data visualisation and exploration'.

Integration

Beyond the advantages offered to archaeology alone, we may also consider the extent to which archaeological databases really constitute a separate knowledge domain. Many heritage records wish to code data regarding the political and/or environmental situation of the sites—whether they are on a particular type of soil, or fall within the bounds of a particular political region.

Spatial technologies offer the opportunity to integrate archaeological data with any other available forms of data that the organisation may have. Many archaeological management contexts have types of data available that have traditionally been stored separately, often with some available in a computerised record and others not. A good example of this is the use of aerial photography, which often exhibits clear evidence of archaeological sites. Where many organisations have both a database of recorded sites and a photographic archive, GIS presents them with the opportunity of holding these within an integrated system: known sites can be displayed against the photographic evidence for their existence and extent, and new records can be created by, for example, 'heads-up digitising' of visible features on rectified photographs.

It has long been recognised that few archaeological managers exist within solely archaeological organisations, and so there is frequently a synergy to be obtained through the integration of various heritage, environmental and political data sources. That to date this has not often been successfully achieved is perhaps due more to institutional factors than intellectual ones:

"Archaeology has been painfully slow to integrate with other areas of landscape conservation despite the fact that the archaeological record is a finite and rapidly disappearing resource." (Lock and Harris 1991:170)

Where it has been successfully achieved (see examples below) then archaeologists add considerable value to their existing data. Maps of soil, geology, political areas and topography are stored within, or are accessible from, the same system as the archaeological information. Quite apart from the obvious gains in efficiency and reduced bureaucracy (*e.g.* archaeologists no longer have to apply to, or visit, another part of the organisation to use their data) the range of questions and outputs available to archaeological managers is multiplied considerably in these circumstances.

Actually, it is one of the supreme ironies of information technologies that many of these problems did not become problems until the advent of computerised records. The 'card-and-map' system is an excellent example of the use of appropriate technology to record the spatial and attribute components of archaeological data. The growth in numbers of sites made the attribute side of this system impossible, but the solution adopted made the spatial component unmanageable.

11.2 ARCHAEOLOGICAL RESOURCE AS CONTINUOUS VARIATION

One of the most obvious characteristics of most computerised records of archaeological heritage is that they are discrete. To understand this, we need only to consider the way that *value* (or *formal status*, if the term value is problematic) is ascribed to elements of the heritage record. Early records regarded 'sites' or 'monuments' as units to which particular status may be attached. In the UK, legislation since the late 19th century has allowed particular sites to be designated as 'Scheduled Ancient Monuments' if they are deemed to be of 'national importance'. Considerable effort has been expended on delimiting and locating the extent of this designation, and equivalent efforts have been made in many other systems worldwide.

It is in support of this conception of heritage 'value' that management-oriented information systems were established, and such systems have adhered closely to this notion of the 'valued' site. As a result, most heritage databases recognise only the existence (or not) of a site or monument. In the most advanced systems, attributes may be recorded about areas as well as points but all, inevitably,

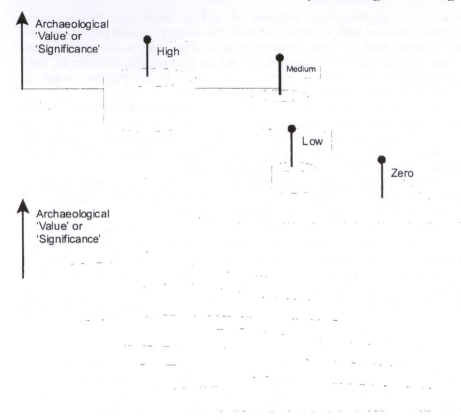

Figure 11.1 Discrete (top) and continuous (bottom) notions of the variation in archaeological value or significance across space.

consider archaeological value to be discrete, something that can either be present (and measured on a categorical or ordinal scale at best) or not.

    However, a little thought suggests that this is not the only way of thinking about archaeological 'value' (or 'importance' or 'significance'). It is well understood that the archaeological 'value' of any location in the landscape is actually far more complex than the various legislative possibilities allow. For example, in the UK the criteria for a known site becoming a Scheduled Ancient Monument are complex and recognise that the national importance (or not) of any given site might depend on the setting of the monument, the density of other monuments around it and other issues that cannot be regarded as discrete. We might also consider the spaces between the discrete locations that archaeological management legislation decrees to be 'important'. It could be argued that any, or all, of the following factors can impact upon the archaeological 'value' of a place:

1. the likelihood that it contains an undiscovered archaeological site;
2. the proximity to known archaeological sites (or to areas likely to contain undiscovered sites);

3. the visibility of known archaeological sites from this place (or the visibility of this place from known sites);
4. the likelihood that this place was habitually passed through in transit between archaeological sites.

This list could go on, but it is clear that if archaeological 'value' is to have any meaning then it is better conceived of as *continuous variation* rather than discrete lumps of value surrounded by a landscape devoid of archaeological importance (Wheatley 1995b). Further, we should probably say *variations* plural, because it is very unlikely that we could formulate a single summary of archaeological value, and even more unlikely that all members of a community would agree on it if we could.

Spatial technologies allow archaeological managers to manipulate representations of space, not just records of archaeological sites. Because of this, many have begun to formulate representations of the cultural heritage that move beyond the simplistic and begin to take account of the continuous nature of the archaeological record. Institutional momentum will mean that in most parts of the world it will be a long time before heritage legislation recognises the whole landscape as archaeological, but that is not to say that heritage managers cannot begin the process themselves.

11.3 REALITY: THE ANTIDOTE TO GIS

This chapter has so far been universally optimistic, possibly even evangelical in tone. However, it is important to recognise that there is a significant difference between the *potential* of spatial technologies to liberate, empower and improve the management of cultural heritage remains and the *reality* of their implementation. In this respect, we must review some of the most significant problems encountered when heritage agencies adopt or plan for technologies such as GIS.

Most organisations approach GIS implementation with the view that the technology is complex and that installing hardware and software followed by migration of existing data sources will present the greatest problems. Experience suggests that a working system connected to (or containing) existing heritage data is usually where the real problems begin.

Data and data quality

The first issue that GIS implementation usually reveals is that of data quality. This might be better phrased as the 'fitness for purpose' of the archaeological data because, often, the data is entirely adequate for the purposes to which it has been put in the past. Where the data are collected or revised specifically for a given purpose, as in the case of the *Dolmen* database discussed below, then formal data quality criteria can be set and quality controlled as the data are collected. However, the vast majority of archaeological databases utilise existing data that may have derived from earlier traditional databases or from paper records. Quality of attribute data is no more of a problem for spatial information systems than in a conventional database application—which is not to say that it presents no difficulties—but the spatial and topological components can present significant problems.

To give an example, Boaz and Uleberg reported that they faced considerable problems in managing Norwegian cultural heritage material because firstly the highly forested nature of some of the areas concerned meant that accurate surveyed locations of sites were uncommon, and secondly because the record as a whole was highly unrepresentative of the overall pattern of sites in the region (Boaz and Uleberg 1993). The first of these is a straightforward issue of precision and comprises one of the main problems that surround the recording of the spatial component within traditional databases. Without visualisation tools that represent the spatial location of a site, these are very often inaccurate to the extent that sites may be located in rivers, lakes or other parts of the world.

Related to this is the issue of precision of site locations. It is not unusual to have a database in which it is not the survey precision that creates the problem but the spatial precision of the grid references used. Many of the procedures used for recording in a text-based system are entirely suitable for an archaeologist to find the site in the computerised record, look up the grid reference on a map and find the corresponding record. However, they may prove to be quite inadequate for use within a computerised mapping or GIS system. Worse than this, it is common for databases to employ varying levels of spatial precision that depend on who recorded the site and when. For example, it is not uncommon for some sites to be recorded to the nearest of 100m, some to 10m and some to the nearest 1m. In many records, the spatial referent in the conventional database does not mean '*this site is at this co-ordinate with a given precision*' but instead means something like '*the site is in the grid square whose lower left hand corner is described by this co-ordinate*'. This is what we mean by 'fitness for purpose' because such conventions are well understood and perfectly serviceable in the context that they were originally intended, but hugely problematic for a system that explicitly codes the spatial and topological components of archaeological data.

Again, this is an issue that has been recognised for some time and widely discussed, although it repeatedly recurs as if it had never been considered before. Characteristically, Diana Murray has put this succinctly:

"Information which has been collected over a long period, mostly with very different purposes in mind, is being stretched to the limit to provide distribution analyses or predictive models. Systems have been designed to pose questions for management and research purposes which could never have been contemplated before." (Murray 1995: 86)

The solutions to these problems are not simple and are certainly not cheap. There is no automatic method for resurveying sites and monuments, although identification and correction using rectified aerial photographs is sometimes possible. Improvements in data quality must inevitably be undertaken by manual intervention, either laboriously in a single 'enhancement' project, or piecemeal over many months, years or decades. The former approach is obviously attractive but rarely practical because of the sheer workload that can be involved in creation or modification of a heritage database. For example, Palumbo reported in 1992 that the coding of data for the JADIS database had required 10 months work by six employees and was still ongoing (Palumbo 1993). He expressed a hope that more staff would be made available, although history does not record whether this plea was ever granted.

People, time and training

This brings up the second area that can bring noble aspirations into violent contact with reality: the ability of archaeological managers to actually undertake any of the new kinds of information management described above. Many have little, or no, formal training in information technology, and none whatsoever in GIS. Faced with an implementation of a high-end vector-GIS, most archaeological managers find that they must re-learn many of their existing technical skills before any of the benefits we have discussed above can be realised.

Closely related to this is the fact that archaeological managers are generally archaeologists with an archaeological job to do. However well resourced a training and support system is, it cannot succeed unless staff are provided with the time to take advantage of these opportunities. We would go so far as to say that time is the most significant issue that holds back organisations and individuals who wish to implement GIS-based heritage management records.

There are many things that can be incorporated into a project design to help in this regard. Formal training and education is essential but often very expensive. Great care also needs to be taken that it is delivering the right skills and knowledge to staff. A good case can be made that individual archaeologists themselves are best placed to decide which skills and knowledge are most immediately relevant, and so self-tuition can therefore be far more effective. In our experience, most archaeological managers are ideally suited to self-tuition because they are generally well motivated and have good independent problem-solving skills. The provision of good instruction manuals and tutorials is therefore essential, and it goes without saying that large numbers of copies of this book should also be provided. Where networked resources are available, then tutorial material and learning resources can be provided and delivered via intranet or internet.

Beyond this, thought can be given to the design of the systems themselves. Ideally these should be implemented with several different technical levels of user in mind, either through different interfaces (as in the HCC example discussed below) or through the provision of entirely different clients that are linked to the same core data.

Planning and project management

The conventional answer to these issues is to plan ever more thoroughly: to identify, cost and manage all aspects of the project from the outset. This can involve the identification of the functional requirement for the system, selection of software and hardware platforms, establishment of a formal procurement procedure, provision of training and many more stages—see for example Lang (1993), Clubb and Lang (1996a, 1996b). There is no doubt that good planning is vital to success, but there is a very real danger of 'paralysis by analysis'—where the process of planning an IT project itself becomes so overwhelmingly complex and bureaucratic that the systems themselves are never successfully implemented (or if they are, they are implemented by committee, which is usually worse). It would be quite unfair to cite specific examples of this, although there are real examples of major heritage institutions that made significant investments in *discussing* and *planning* for GIS implementation over more than a decade without ever implementing the systems concerned. These have even included—with apparently no appreciation of irony—published discussions of the role of 'post implementation reviews'. Many archaeologists in national or regional heritage management organisations will no doubt be able to think of examples.

Case studies

To illustrate the ways that spatial technologies are currently being deployed within archaeological management we will take a careful look at two rather different case studies. The first concerns a relatively small-scale system—although managing a substantial body of data—that has been designed to fulfil a highly specific user need and to provide access to a tightly defined range of spatial and attribute data. The Forest Enterprise (Wales) *Dolmen* GIS and database have been implemented using relatively inexpensive single-user software systems, and is an excellent representative of this type of application. In contrast, the systems used by Hampshire County Council are required to serve a large number of employees, both archaeologists and others, whose needs vary considerably. The systems are designed to integrate a wide variety of data, and deliver this in support of users who have not only different needs, but also widely different expertise in the use of information technologies.

11.4 SEEING THE WOOD FOR THE TREES: *DOLMEN* DATABASE AND GIS

The first case study we will describe is the *Dolmen* system of Forest Enterprise (Wales), a part of the Forestry Commission of Great Britain and responsible for managing about 120,000 hectares of Welsh Forestry Estates. Cultural resource management on this land presents particular archaeological problems: heavy machinery, used during harvesting, can easily damage unmarked archaeological remains, and the level of knowledge and recording of many archaeological sites has been limited since the planting of managed forests in the late 1940s.

The context of the application

In recognition of this, Forest Enterprise proposed and carried out the *Welsh Heritage Assets Project* to establish a comprehensive record of the cultural heritage resources on the Welsh Forestry Estates[7]. Data acquisition was subcontracted to a variety of different organisations, which were required to meet strict data quality criteria. The task involved consulting existing Sites and Monuments Records (maintained by local government) and digitising locations of known sites from mapping that pre-dated the establishment of the forests. Previously unrecorded sites were also located and the archaeological data were both verified and extended

---

[7] The project was initially proposed by Clwyd-Powys Archaeological Trust and a pilot project was funded by Forest Enterprises. The concept was then developed by Forest Enterprises and Caroline Earwood between 1995 and 2000.

through site visits. Obviously, one of the principal concerns of the project was the quality and management of this substantial database, and to this end both the quality assurance and the development of the software tools to utilise the database were contracted to the *GeoData Institute* of the University of Southampton. The data collected by the project was subsequently returned to the regional records and so serves the dual function of providing a resource to Forest Enterprise staff, as well as providing an enhancement to regional Sites and Monuments Records.

The main role of the *Dolmen* GIS[8] and database system is to provide information to the five Forest Districts to support archaeological and conservation issues. The information therefore has to be delivered to a highly specialised group of users each of whom requires a subset of the database relevant to their particular district. Throughout the project, Forest Enterprise staff played a critical role in providing information about their needs as well as feedback during the trial period.

In addition to the archaeological data and attributes (such as type, period, condition and management category), the system needed to provide the user with access to a range of other geographic data, including hydrology, land boundaries, trackways and forestry information. Functional requirements of the system included relatively conventional functions for report-writing and accounting for expenditure, but also provision of access to photographs taken during site visits (stored on CD-ROM) and the ability to undertake some basic spatial operations such as query the database on spatial criteria and generate buffers around archaeological features. When felling is planned, the system is used to decide, for example, the optimal route for movement of machinery into and out of the area without damage to archaeological remains.

Design of the system

As a result of these requirements, the system comprises a standalone product consisting of an integrated database and GIS. The database management system is implemented as a Microsoft Access application, linked to a customised ArcView interface, delivered as an ArcView extension. The database system can be used independently of the GIS to provide access to the data, undertake financial management and to generate standard reports. The implementation of the GIS as a customised interface of an existing product, provides all of the necessary functionality for planning related to felling *etc.*

To manage the acquisition of new data, without compromising the exchange of information between Forest Enterprise staff and regional Sites and Monuments Records, the *Dolmen* system incorporates a distinction between conventional archaeological sites and 'local' data. Local data can be used to record additional archaeological information, or to create records of newly discovered sites. These are then passed back to the regional archive for checking by archaeologists prior to incorporation into the main dataset. Rather than link the various *Dolmen* databases

---

[8] The design for Dolmen was prepared by Dr Caroline Earwood, project manager of the Welsh Heritage Assets Project and it was then created by the GeoData Institute, University of Southampton. The WHA Project was part funded by the Heritage Lottery Fund, although this did not include the development of Dolmen, which was completely funded by Forest Enterprise (Wales).

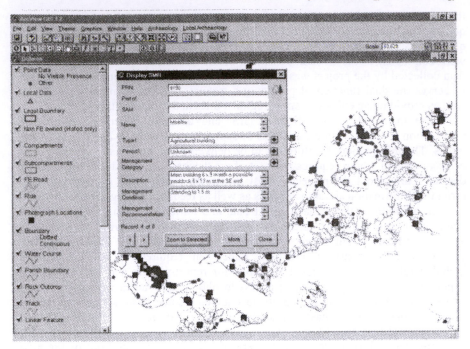

Figure 11.2 *Dolmen* GIS, showing the integration of archaeological, boundary and forestry data within a customised ArcView application. © Forestry Enterprises (Wales). Reproduced with permission.

directly to a central database, the non-local parts of each of the databases is periodically overwritten with fresh data drawn from the SMR.

Discussion

The management structure of this particular project has ensured that data quality issues are unusually prominent for an archaeological application. In this case data quality was carefully managed through the contractual structure, with the contractors required to meet very specific quality criteria at every stage of the project. Within the context of UK archaeology, this can be contrasted with the situation of the regional Sites and Monuments Records, for whom enhancement or management of quality in this way is very rarely possible within the constraints of local government finance. In this respect, one of the most interesting aspects of this example is the way that it creates a synergy between the forester's and the regional records.

　　Dolmen represents a robust and flexible solution to an increasingly common requirement within archaeological management: the delivery of appropriate archaeological and other spatial information to a relatively small group of people who have a clearly defined task. The solution adopted makes maximum use of the existing functionality of widely available software packages to meet the needs of the users. At the same time, the use of a familiar interface and the provision of good documentation significantly reduce the need for formal training, particularly where the users are already computer literate. Overwriting the distributed copies with fresh data periodically is a pragmatic solution to ensuring that data integrity is maintained, obviating the need for complex record locks or other integrity checks, and the overall result is an extremely useful tool.

11.5 REGIONAL HERITAGE MANAGEMENT: HAMPSHIRE COUNTY COUNCIL

"GIS is enabling us to achieve things that we've never been able to do before, like get information to the districts: to bridge the geographic gap between administrators." (Ian Wykes, Senior Archaeologist, Hampshire County Council)

Hampshire is a County in southern England whose regional authority is Hampshire County Council (HCC). This manages, for example, Education Services, Fire and Rescue, Libraries, Road Maintenance and Social Services for the region and is controlled by locally elected councillors. Within the HCC organisation is a *Planning Group* who are responsible for, among other things, providing advice on planning issues to other planning authorities within the Hampshire region.

In support of this function HCC, like most other regional authorities, maintains records of the archaeological sites and of protected buildings within the region and use these to provide advice when an application for development is submitted to one of the Local Planning Authorities within Hampshire. Hampshire is a useful example to us, because the systems that they have implemented are unusually sophisticated and coherent. As such, a consideration of the purpose and utility of their systems has wide relevance to other similar regional heritage management contexts both in the UK and elsewhere.

Historical background

Like most of the regional Sites and Monuments Records (SMRs) within the UK, the origins of Hampshire's archaeological database lie in the first national archaeological record created by the national mapping agency for the UK, the Ordnance Survey, in the 1920s and 1930s. In 1976, the relevant parts of this record became the first Hampshire SMR. This was a paper-and-map system, in which archaeological site details were recorded on numbered cards, and the numbers were also recorded on paper copies of maps to show the spatial component of the record. Historic buildings records also similarly began as paper records, in this case as published lists of buildings.

These records were computerised in 1981 and both the content and the computer systems used for this computerised record have been enhanced several times. Over the years the data has been held on systems designed 'in house' and others using data management products provided, or designed by, national heritage agencies. Since 2000—and unusually for a regional record in the UK—the computerised records of archaeological sites and historic buildings have been combined within one unified Archaeology and Historic Buildings Record (AHBR), now stored as a relational database (see Figure 11.3).

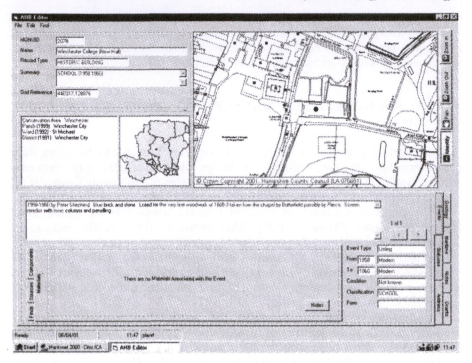

Figure 11.3 Example screen from the AHBR programme, showing integration of spatial and attribute data for a listed building. © Hampshire County Council. Reproduced with permission.

Like any large organisation with responsibilities for planning and environmental management, HCC has a section responsible for providing mapping services, and this holds spatial data such as maps and aerial photographs. Recently these have been available to the council officers as digitised (raster data) as well as in paper form.

Description of current systems

Over a number of years the Planning Group at HCC have developed a highly integrated approach to the management of their data. Since the mid 1990s, the principal tool for integration and interaction with the data holdings has been a spatial technology. The AHBR database is currently implemented as a relational database system, designed by the Planning Group themselves and implemented as an Oracle database server, made available via ethernet connections to clients in the various departments. Because the AHBR implementation involves a relatively major relational database system, the design of client software benefits from the advanced features of such systems such as file locking, transaction processing *etc.* and the availability of tools to manage data security and access permissions.

Since 2001, the AHBR database has been spatially enabled with Oracle's 'Spatial Data Extension', which extends the capabilities of Oracle to those of a full spatial database and permits clients to access data via the ESRI 'Map Objects' interface.

Several different client systems are provided, of which the most widely used was written specifically for the group. The AHBR program now contains all the functionality that was previously provided within a 'forms' interface to the attribute data, but with the addition of a range of spatial data management tools and a graphical display. This provides suitably protected access to other spatially enabled datasets within the organisation, including Ordnance Survey map data at scales between 1:50,000 and 1:2,500, postcode (zipcode) data and data belonging to other groups within HCC.

Where necessary, Arc Info and ArcView GIS products (ESRI) are also deployed as more powerful tools to manage, integrate and analyse spatial information. This also permits the creation, editing and use of other discrete datasets that have been collected for specific purposes, often by contractors working for the group. One of these relates to Historic Towns, which serves to record the towns in the region as area features, along with a number of important characteristics including their surviving defences, church plots and other areas of known or suspected archaeological potential.

Data quality issues

As with many sites and monuments records, the Hampshire archaeological and buildings databases themselves store all records as point data. Presently, none of the spatial extents of the sites are digitised or surveyed although there is now no technical reason why this cannot be done. However, an idea of the scale of the undertaking that would be required to enhance the database in this way can be gained by considering a far more minor enhancement of the historic buildings data that was undertaken in 1998.

With the unification of archaeological and historic buildings data, the Planning Group decided that location data for the 14,500 historic buildings records was not sufficiently precise. A very high percentage of the records were proving to have grid references that fell within the mapped boundaries of the wrong building, the middle of the street, within rivers or elsewhere. To rectify this required a full time employee for 18 months to check each record, with the result that currently it is estimated that 90-95% of records are accurate to the extent that the points fall within the mapped outline of the correct building.

People issues

On average, archaeologists at Hampshire spend around 1 full day attending formal training with the majority of training needs being met in informal ways. Of vital importance here is the implementation of several client systems with different complexities and capabilities. Archaeologists with no need for sophisticated analytical procedures are provided with the basic tools necessary for information management by the in-house client. In contrast, those who need more flexibility or a wider range of analytical functions can use ArcView, which is connected to the same data. The IT manager and those with a need to manipulate and maintain the databases themselves are provided with a powerful and fully topological GIS in the form of Arc Info.

This approach to the implementation of systems considerably 'shallows' the learning curve for new users of spatial technologies, and has been so successful that the group are now planning to implement a simple spatial query interface for WWW browsers. This will provide even less functionality than the in-house client, but is likely to achieve two things. Firstly, it will likely serve the needs of occasional users of the database systems fairly completely, freeing them from the need to learn even the relatively straightforward interface of the in-house client. Secondly, it will provide another shallowing of the learning curve of the spatial systems, and should contribute to the already considerable pool of skills that has developed within the group.

11.6 NATIONAL AND SUPRA-NATIONAL CONTEXTS

The use of spatial technology to manage archaeological inventories is necessarily defined by the institutional context of that management, and in this respect we might, albeit rather arbitrarily, consider a 'national' context for heritage management. This *is* rather arbitrary because there is tremendous variability between nation states as to the extent to which cultural resource management is devolved to regions. Many modern states had highly devolved systems of heritage management, for example Spain, where the management and recording of cultural resources reflects the highly devolved nature of government (see *e.g.* Blasco *et al.* 1996, Amores *et al.* 1999). Other nation states, such as France, have been characterised by far more centralised approaches (see *e.g.* Guillot and Leroy 1995). More commonly there are both national and regional systems in place, with varying degrees of responsibility for cultural heritage. Examples include the United States, where regional (State) services maintain a significant responsibility for management of heritage, with Federal agencies (forest and national parks agencies) holding responsibility for substantial tracts of land (*e.g.* Hutt *et al.* 1992). Another example is the United Kingdom, where regional (County) records are used in support of archaeological management through the planning system, while national heritage agencies (English Heritage) and other national bodies (National Trust) have rather different responsibilities, and therefore different requirements from their systems.

A survey of the level of adoption of spatial systems for managing national and regional records within the European Union was recently conducted by Garcia-Sanjuan and Wheatley (1999). This found, perhaps unsurprisingly, that these variations in centralised/regionalised traditions of heritage management presented considerable obstacles to the successful implementation of GIS. Interestingly, they also found a relationship between the date that computerisation of heritage records began and the implementation of spatial technologies for handling those records— a relationship that suggested that early adopters of database management systems were *less* likely to be early implementers of spatial technologies. The reasons for this are complex, but are likely to be partly related to the problems of data quality that we have discussed above.

That there is also now a growing need for some strategic thought with respect to the integration of national and regional inventories is stressed by the growing number of international accords that are in current circulation. For example, the establishment (albeit with no real statutory meaning) of *World Heritage Sites* by the International Council of Museums (ICOMOS) and, in the context of Europe, the ratification of the Valletta Convention on the Protection of the Archaeological Heritage. The latter calls upon the individual states and the entire European Community of Nations to institute and implement administrative and scientific procedures to provide satisfactory protection to the archaeological heritage (COE 1997).

How the integration, or interoperability, of such a diverse series of administrative systems and information technologies can be achieved is, at present, entirely unclear although there should certainly be a role for *metadata* (as discussed in Chapter 3).

11.7 CONCLUSIONS: RECOMMENDATIONS FOR THE ADOPTION OF GIS

In our experience, there is nothing more calculated to enrage and infuriate archaeological resource managers than academics trying to tell them how to do their jobs. Consequently, it is with extreme trepidation that we include some guidelines for professional archaeologists or heritage management organisations who are considering adopting spatial technologies. Nonetheless, our own experiences of professional archaeology coupled with frequent —and sometimes painful—discussions with many colleagues engaged in the management and support of heritage records suggest the following broad points:

• *Planning and project management* are vital to success but can be overdone. The management of a successful IT implementation project requires a combination of technical and people skills that are often difficult to obtain and expensive, and the management procedures themselves need to be flexible enough to allow the archaeologists to 'discover' the way forward.

• *Consideration of context* involves not only the local context and the immediate needs of the users, but also the ways in which the information system could, or should, be integrated within a larger information system. For regional cases, this may involve considering how the system relates horizontally to other regions, and vertically—both upwards, to national systems, and downwards to sub-regions or other organisations. Even national inventories should now consider the future integration or interoperability of systems in a supra-national context.

• *Development of systems* is a job for IT professionals. Beyond the most basic customisation of commercial products, in-house development is likely to succeed only where a full-time professional systems manager is available. As with good project managers, this is also likely to be expensive. The in-house development of bespoke systems may be extremely successful in larger organisations (see the HCC example) but 'homemade systems' can be disastrous if created by poorly trained, inexperienced staff who have other responsibilities.

• *Investment in data* is more financially significant than any software or hardware system. Moreover, the quality of the data will serve an organisation for decades while any technical solution to its management is likely to be replaced on a far shorter cycle. Spatial technology can be a fine environment in which to improve data quality, but it does not do it automatically.

• *Investment in staff* needs to be considered as important as the investment in hardware, software and data. Effective staff development requires careful thought in relation to enhancement projects, and needs to be costed into budgets and provisioned in terms of time. Where enhancement schemes succeed, they are often because a highly motivated team of individuals invest considerable time in the technology and because management is flexible enough to allow them to do this.

CHAPTER TWELVE

Future directions

"Geographic Information Systems have nothing to offer. Their time is past and the 'Third Wave' of computing has broken." (Allinson 1994)

Those who have been predicting the demise of spatial technologies for some years must have been rather disappointed. This is mainly because they have failed to take into account the ability of technology to evolve and develop: just as they seem to offer no further scope for development, spatial technologies like GIS seem to 'morph' into something new and a whole new research landscape opens up for us. In this chapter, we will explore some of the newer areas of spatial technology that may continue to confound the pessimists and delight those who wish to develop new insights into the spatial structure of cultural remains.

As the alert reader will have no doubt realised, however, we do not place a great deal of store in the act of prediction. We argued in Chapter 8, predicting the past is ferociously difficult and it follows that predicting the future is almost impossible. As a result, if there is a single chapter that is guaranteed to have in-built redundancy in a text such as this, it is the 'future directions' section. Rather than being respected in future years for its remarkable foresight, the usual role of such chapters is as cautionary lessons in just *how* wrong a set of the authors *could* be. We digress. In seeking to predict the future, the options available seem to fall into two camps. The first is to effectively side-step the issue by making frequent recourse to another emergent technology (ideally a technology whose state of emergence is not quite as complete as the one you are discussing). Popular examples at present appear to be Virtual Reality (VR) modelling (*e.g.* Gillings and Goodrick 1996) and Artificial Intelligence (*e.g.* Spikins 1995). The second is to grasp the nettle of prediction but try to ensure that you do not lose sight of land by attempting to peer too far. The latter is the approach we have adopted in the following chapter.

In looking ahead we have decided to focus our crystal ball on two distinct areas—trends within the application of GIS to archaeology and methodological advances. In each case we have attempted to root our predictions firmly within current research trends within the sub-discipline.

12.1 THE CURRENT STATE OF GIS APPLICATIONS WITHIN ARCHAEOLOGY

Before attempting to chart trends in the ways in which GIS can be incorporated into archaeological research, it is useful to firstly characterise how it has been utilised to date. A suggested breakdown is given in Figure 12.1, although it should be noted that the rather tidy structure presented is largely for heuristic purposes. As with so many areas of life, whilst neat categories and clear dichotomies are conceptually clean, the reality of the situation is invariably much fuzzier. This is a point we will return to shortly.

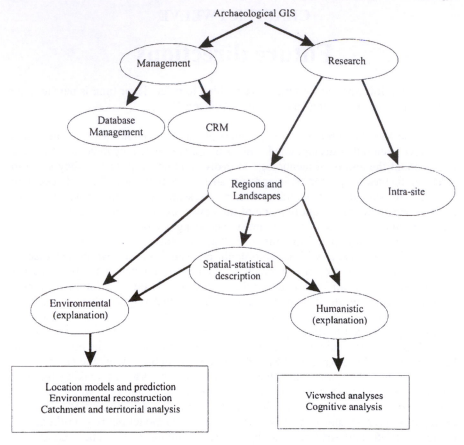

Figure 12.1 A suggested structure for the current application of GIS within archaeology.

For practical purposes, current applications of GIS within archaeology can be seen to fall into two broad areas—these we have termed *Management* and *Research*. Within each of these categories further sub-divisions are evident. Let us begin by taking a look at the Management category. Here we can see a clear split into applications which focus upon the storage, maintenance and analysis of existing National and Regional Databases (what we have termed *Database Management*), and applications which stress the more active management of the archaeological resource (*CRM*). Looking to the former, these applications effectively use GIS to provide a two-way spatial window into an existing database of archaeological information, allowing users to interactively query and interrogate the data set. In this sense the GIS is being used to enhance existing database systems through the articulation of the spatial dimension. The latter type of application is concerned with the active management and protection of the archaeological resource, and involves fields such as development planning and predictive modelling. Here the GIS is being used in a much more dynamic and creative way to aid often severely under-resourced archaeologists in devising approaches to best protect and maintain the cultural resource. For example, we have already argued in Chapter 11 that spatial technologies provide an opportunity for managers to implement a more continuous notion of the archaeological resource.

Looking now to research, once again we can identify a clear split into applications focused at the inter-site, regional level (*Regional Landscape*) and specific excavation and site-based studies (*Intra-site*). In highlighting this dichotomy it is important to acknowledge that the research arena has been almost totally dominated by only one of these trajectories: regional landscape-based studies. It is interesting to note that this bias has been a characteristic of archaeological-GIS applications from the outset. This is perhaps due to the over-riding landscape emphasis of the first wave of archaeological texts dealing with GIS. For example, the first major textbook focusing explicitly upon the relationship between archaeology and GIS was published in 1990. Entitled *Interpreting Space: GIS and archaeology* (Allen *et al.* 1990), it is of particular interest in that *all twelve* case-studies described in this ground-breaking volume had a regional landscape focus. In the introduction it was even asserted that *only* landscape-based archaeological research could provide the conceptual framework necessary to take full advantage of the potential the GIS methodology had to offer (Green 1990). In 1991 the first explicitly European GIS case-study appeared in print, Gaffney and Stancic's Hvar volume (Gaffney and Stancic 1991). Here the application of GIS within the established context of regional survey was articulated even more explicitly. Despite the introductory nature of many of the analyses held within, this latter account has remained one of the most influential and undoubtedly accessible archaeological introductions to GIS. What seems clear is that in the early 1990s a direct equation between GIS and regional landscape-based study became sedimented within archaeological practice, an assumption that still exists as a powerful bias today.

Although a number of intra-site studies have been attempted, these have been very much in the minority. Such studies are severely hampered by a combination of two fundamental shortcomings of existing GIS systems (Harris and Lock 1995:355-357, Harris and Lock 1996:307), coupled with marked variability in the source data employed. As both of the methodological limitations will be discussed in more detail in the second part of this chapter we will only outline them here. The first concerns the fact that whilst excavation recording is invariably undertaken in 3-dimensions (3D), the routinely available GIS packages used by archaeologists are unable to handle truly 3D information, *i.e.* information where attributes can be recorded for any unique combination of 3D space. Where intra-site applications have been successful it has tended to be in the context of studies where the stratigraphic variability of the source data is limited (Vullo *et al.* 1999).

A further fundamental problem is the lack of a truly temporal dimension to GIS. We might think of excavated data (if not all archaeological information) as varying in potentially 4 dimensions. Existing GIS can handle 2D space very well but can only handle chronological information (what we might think of as the biography or history of things and objects) as an attribute. The result is that if temporal dynamics are to be explored in a given study, the only option available to researchers to is to employ highly simplistic, snapshot-based strategies such as time-slices (Biswell *et al.* 1995, Mytum 1996).

However, the lack of effective intra-site studies is not solely the result of inherent limitations in the GIS. There is another problem related to the quality of data collected in the context of modern excavation practice. The quality and resolution of the data is often not consistent or of a quality to facilitate detailed spatial analysis. Although the excavation and recording may have been up to accepted modern standards, it is only when an attempt is made to articulate the spatial facet within a GIS that inconsistencies are highlighted. These can refer to variations in the spatial referencing of objects and the resolution/recording of certain contexts. Such inadequacies have been noted by a number of researchers ranging from those involved in the analysis of data yielded by traditional excavation, for example Biswell *et al.* (1995:283) at Shepton Mallet, through to those involved in the study and identification of sub-surface deposits such as Miller's studies in York (Miller 1996). Taken together these limitations have served to greatly inhibit the broader application of GIS within the intra-site context.

Let us return to the most common application area, the regional landscape context. Here an answer to the question 'what are landscape-based archaeologists doing with GIS?' could be answered in the main by the following list of options:

- rediscovering (and invariably proselytising about) the quantitative spatial statistical approaches characteristic of the 'New' and 'Spatial' Archaeologies of the 1960s and 70s;
- predictive modelling—privileging environmental information;
- 'humanistic' analyses—privileging cultural information.

As a result we have broken the Regional Landscape context down into a number of distinct sub-classes. These are differentiated largely on the basis of the interpretative emphasis that is placed upon the particular type of source data used to furnish explanations—ecological or social. In each case a boxed list of typical analyses is given to better illustrate the types of study we are placing within each category. On the one hand we have *Environmental* explanatory approaches. These refer to more ecological approaches to explanation, whether concerned with modelling post-depositional change or using deterministic models to approach and describe patterning in the archaeological record. In contrast, *Humanistic* explanatory approaches are those that place a premium upon social and cultural variables. In each case the explanation may be preceded by a descriptive statistical analysis—in effect identifying that there is something to explain in the first place.

As intimated earlier, the snapshot of GIS applications presented above should be treated as a simplification of a much more complex reality. For example, the clean split between Research and Management is very difficult to maintain, as the two strands invariably inform one another in any given analysis. In addition, the stark dichotomy identified between 'environmental' and 'humanistic' approaches may be as much a reflection of the rhetorical positions adopted by researchers—see for examples Wheatley (1993, 1998, 2000), Gaffney and van Leusen (1995), Gillings and Goodrick (1996), Kvamme (1997), Wise (2000)—in response to far wider theoretical debates, than the reality of everyday GIS-based analyses.

In another recent review of the history of GIS applications in archaeology, Aldenderfer stated that:

"It is important to stress, however, that as a tool, GIS and associated technologies are 'theory-free', in that there is no necessary isomorphism between a particular data type or category and the use of GIS to solve or explore a problem." (Aldenderfer 1996:17)

We would not endorse the notion that any tool is 'theory free' in any simple sense—see *e.g.* Zubrow (1990b), Wheatley (1993, 2000) or (Gillings 2000), for a discussion of the inherent dangers of the notion of GIS as 'neutral tool'—although Aldenderfer's broader point is a good one and provides a useful point of departure for our own (admittedly rather utopian) predictions of the future shape of GIS applications within archaeology.

12.2 THE DEVELOPING SHAPE OF GIS APPLICATIONS WITHIN ARCHAEOLOGY

One of the clearest trends in the application of GIS within archaeology is likely to be the gradual collapse of these existing dichotomies, most notably that between 'environmental' and 'humanistic' approaches. This is partly a reflection on the futility of continuing to adopt intractable positions, but is also a reflection of the wider theoretical concerns with deconstructing the culture:nature dichotomy that the debate is predicated upon. Instead, a more judicious and critical blend of environmental and cultural factors may begin to inform and characterise spatial analyses. The key factor that will make this possible will be the explicit acknowledgement and integration of developments in archaeological theory into GIS-based studies.

One early, and frequently repeated, criticism of GIS was that the analyses undertaken displayed a profound lack of imagination. This was rather elegantly characterised by Harris and Lock as a form of *technological-determinism* (Harris and Lock 1995:355). In effect, the GIS enabled researchers to routinely undertake certain types of analysis. As a result researchers either undertook such analyses directly, or fished around for a suitable theoretical/conceptual framework within which they could be accommodated (Wansleeben and Verhart 1995). A good example can be seen in the close similarity which exists between the routine analytical calculations of buffer generation and map overlay with the methodological pre-requisites of Site Catchment Analysis. As a number of researchers have acknowledged, many of these re-discovered and appropriated techniques had fallen out of favour for very good reasons, yet the re-emergence of interest heralded by GIS came at the expense of any awareness of the critical literature that accompanied them.

Instead, applications of GIS must be carefully shaped around specific archaeological questions, themselves embedded in an explicit body of archaeological theory. That this is currently being realised across both sides of what has traditionally been seen as a rather intractable Environmental: Humanistic divide, suggests that such a goal is readily achievable—see *e.g.* Church *et al.* (2000) with respect to predictive modelling; Gillings (1998) on environmental reconstruction and Wheatley (1995a) with respect to viewshed analysis.

Such an embedded, theoretically and archaeologically informed GIS will not only lead to much better archaeology but will also result in GIS-based studies making their own contribution to the development of archaeological theory and practice. To return to the example given above, whilst the uncritical appropriation of Site Catchment Analysis as a methodological approach may offer little new to archaeology as a discipline, one of the by-products of attempts to employ GIS to undertake catchment studies has been the introduction of the cost surface, a heuristic with considerable potential utility. In addition, recent work into viewshed-based analyses have the potential to contribute to recent theoretical debates regarding experience and perception in the development of peoples understandings of their life-worlds (*e.g.* Llobera 1996, Loots *et al.* 1999, Wheatley and Gillings 2000).

12.3 TECHNOLOGICAL DEVELOPMENT OF GIS

As was intimated in the discussion of intra-site applications, one of the major problems with applying GIS to archaeological data concerns intrinsic limitations with the data models employed. Not only do archaeological features and spatial phenomena have to be translated into either a vector primitive or cells in a raster grid, but each of these data models is profoundly 2-dimensional. Fortunately one of the clearest benefits that has resulted from the ubiquity of GIS in virtually all spheres of academic and commercial life is a very dynamic developmental trajectory. There are three distinct areas in the ongoing programmes of enhancement and development of GIS that we believe will have an important impact upon archaeological research. They are:

- Object-oriented (OO-GIS)
- Multi-dimensional GIS (3D GIS)
- Temporal GIS (TGIS)

It is important to note that each of these either already exists in commercial form (3D-GIS; OO-GIS)—albeit not in widespread archaeological usage as a result of factors such as cost—or represents a core area of mainstream GIS research (TGIS). In the following sections our aim will be to outline the key aspects of each of these developments, highlighting the potential each has for an enriched set of archaeological approaches.

12.4 OBJECT-ORIENTED GIS (OO-GIS)

As discussed in detail in Chapter 2, the most commonly used spatial data models in archaeology are vector and raster. These serve, in effect, to translate between the real world spatial information we seek to record, manage and analyse and the GIS-based spatial database. To use these models we invariably have to translate the features of interest into a form suitable for the specific data model selected. Typically we parcel the world up into a series of thematic layers and then represent either the spatial features within each layer as points, lines or areas (vector) or map the behaviour of a given attribute over space as a series of cell values in a regular grid (raster). As discussed in Chapter 2, this process often involves a considerable degree of simplification and compromise, as many sources of archaeological data do not fit naturally, or comfortably, into either of the data models.

Object-Oriented GIS seeks to overcome this problem of abstraction by offering a new approach to the structuring of geographical space. This is an approach which seeks to view the world not as a series of discrete features, or attributes varying across space, but instead as a series of *objects*. For example, to

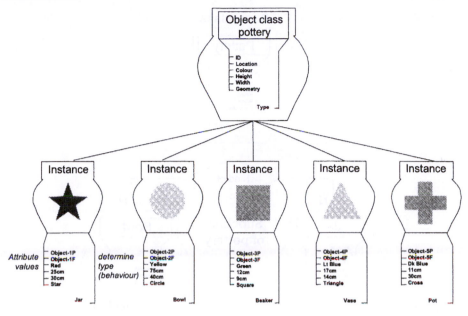

Figure 12.2 'Form is temporary, but class is permanent' (anon). An illustration of a series of instances of the overall Pottery class. Redrawn from Tschan (1999) with permission.

the archaeologist a certain class of prehistoric enclosure may be described along the lines of 'a sub-rectangular ditch and bank enclosing a number of pits, hearths and post-built structures'. It is an odd archaeologist indeed who would describe them instead as either polygon features with an associated database of attributes (number of pits, number of hearths *etc.*), or a cluster of grid cells recording the attribute 'presence of an enclosure'. OO-GIS seeks to model the real world in much the same way as we routinely describe and understand it—in an OO-GIS, the enclosure would be recorded as a class of object: an enclosure.

Class is an important concept in OO-GIS. A class can be thought of as a template that describes the structure of the objects held within it. For example, we may want to establish a class called 'prehistoric enclosure' which we know always comprises a sub-rectangular ditch and bank enclosing a variable number of hearths, pits and post-built structures. An object instance is quite literally an instance of the class—in this case a specific occurrence of prehistoric enclosure. The instance uses the class template to define its specific attributes, each of which will have instance specific values—in this case area enclosed; number of hearths; pits and buildings. In a recent discussion of OO-GIS in archaeology Tschan has illustrated this concept using the example of pottery finds (Figure 12.2).

A further key concept in OO-GIS is that of *inheritance*. Within an OO-GIS as well as general classes it is possible to have sub-classes. Taking once again the object class Pottery as an example, we may in turn want to specify a number of what might be thought of as sub-classes *e.g.* jar and bowl, which themselves represent specific object types (Tschan 1999:312-313). These can be thought of as descendants of the general class pottery and as such automatically inherit all of the traits from the parent class (Pottery), as well as adding specialised attributes of

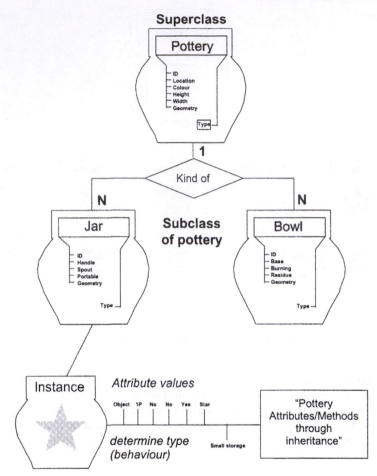

Figure 12.3 Classes, sub-classes and inheritance. Redrawn from Tschan (1999)
Figure 5, with permission.

their own. In this way very complex hierarchies of objects can be established that very closely, and intuitively, model the real world.

Let us illustrate with an example. A bell-barrow is a distinctive type of burial mound found in Bronze Age Britain and we may be interested in exploring the locational factors influencing the siting of such features within the landscape. In a raster-GIS we would create a layer called 'site-presence' and encode each cell in the grid that contains a bell-barrow with a positive value and all remaining cells with zero. We could equally create a vector layer and then decide whether to represent each bell-barrow as a point or an area feature that can be linked to a database of attribute information. In an OO-GIS we would instead define a class of objects termed 'bronze age burial mound' that has a specific set of attributes: area; height; volume; date; number of burials. We could then specify a series of sub-classes: bell-barrow; round-barrow; dish-barrow; pond-barrow *etc.* that inherit all of the attributes of the parent as well as adding their own unique attribute– shape.

The topic of OO-GIS is a complex one and it is clear that in a short discussion such as this we are unable to do adequate justice to the full conceptual and methodological details involved. Readers interested in investigating OO-GIS more fully are referred to a number of recent publications on the topic *e.g.* Tschan (1999)—which includes a useful summary of potential advantages/disadvantages to archaeology—and Worboys (1995). What is clear is that OO-GIS offers a number of exciting possibilities for the way in which archaeologists can model the archaeological record in an information system. Commercial OO-GIS systems currently exist and while at present the high-cost has resulted in very limited archaeological exposure, as OO-GIS becomes more accessible archaeologists should certainly be encouraged to explore its potential.

12.5 MULTI-DIMENSIONAL GIS (3D-GIS)

As a number of researchers have noted, all of the information that archaeologists meticulously record and analyse is profoundly 3D in character. Yet the GIS packages in routine use by archaeologists rely upon a 2D abstraction of reality (Harris and Lock 1996:355-357). In existing systems we can effect a compromise by recording the third dimension (elevation) as an attribute for any unique location recorded along the two axes of 2D space. However, in such systems, we encounter problems if we have two points with exactly the same X, Y co-ordinate position and yet different heights. Although they clearly reference different spatial locations they would be indistinguishable to the GIS—effectively the same location. In a true 3D system any given location would be free to vary along the three independent axes of 3D space. Thus our two points would be recognised as occupying different spatial locations.

As with OO-GIS, commercial 3D-GIS systems do exist but they are frequently very expensive and tailored to specific application areas such as geological modelling and petroleum/gas exploration. As a result they have not witnessed widespread application in archaeological research. One of the most detailed archaeological discussions of 3D-GIS has been that of Harris and Lock, who reviewed in detail the principal methodologies and data models involved in the context of applications of GIS to excavated archaeological data.

Here they employed a *voxel*-based approach to the representation and integration of 3D data. A voxel is best thought of as the 3D equivalent of a 2D pixel (Worboys 1995:317). It can be defined as a rectangular cube which is bounded by eight grid nodes. In a similar fashion to traditional raster data structures, the voxels can be held as a 3D array of voxel centroids with associated attribute data, or as an array which describes the region of space occupied by a given object (Harris and Lock 1996:309). Although the voxel structure suffers from the same problems as 2D raster-GIS—cell resolution effects, an inability to represent precise spatial boundaries, and a lack of topological information—in a feasibility application based upon the 3D interpolation and analysis of borehole data taken around a Romano-British settlement, it produced very encouraging results (ibid:311-312).

As with OO-GIS, multi-dimensional systems have the potential to not only rectify the current bias away from intra-site studies, but also greatly enrich existing landscape-based studies. For more detailed discussions of the various data models and structures involved in multi-dimensional GIS, readers are referred to Raper (1989) and Worboys (1995).

12.6 TEMPORAL GIS (TGIS)

"While it may well be that a geographer's spatial and temporal dimensions do interact... spatiotemporal data are more generally useful when space and time are recorded separately." (Langran 1992:29)

"I do not want to think about time in any philosophical or metaphysical sense. I am searching for operational concepts of time-concepts that will lead to its direct measurement, to better theory, and a richer set of models using time as a variable." (Isard quoted in Langran 1992:28)

At present, temporal information is poorly catered for in existing GIS systems. Most commonly, temporal information is integrated as an attribute and used to generate time-slices—static layers showing a given situation at a fixed point in time. The most common example of this practice is the generation of phased distribution plots.

Although a considerable amount of mainstream GIS research has been directed towards investigating the possibilities of TGIS, the impact the fruits of these initiatives have had on archaeology has been limited. The reasons for this have been neatly summarised by Daly and Lock (1999). Firstly, much of the theoretical research into TGIS has not been translated into usable, functional GIS modules. Secondly, GIS research into temporality has been directed by agencies who themselves have a very specific, modern, western understanding of what the time they seek to model actually is. This is the time of commerce and administration—linear, sequentially ordered and measurable in units of years, hours, minutes *etc.*—clock time. As a discipline whose subject matter is thoroughly temporal, it is perhaps surprising that full and explicit discussions of time within the disciplines of archaeology and anthropology are a relatively recent phenomenon (Shanks and Tilley 1987, Gell 1992, Adam 1994, Gosden 1994, Thomas 1996). However, what is emerging from such studies is a growing realisation as to the sheer complexity of social understandings of temporality and the ways in which space and time are deeply intertwined and implicated within one another. Clock-time is but one understanding and one that may be profoundly tied to capitalism and the modern west. To return to the quotes at the start of this section, the key point we wish to emphasise is that archaeologists are beginning to acknowledge the importance of precisely the philosophical and metaphysical concepts of time that the researchers involved in TGIS have been so keen to ignore.

The issue of integrating time into the GIS is clearly one where archaeologists can make a significant contribution to the discipline of GIS as a whole. Indeed, a number of researchers have already begun to explore the potential of existing raster-GIS for identifying and quantifying temporal concepts such as 'change' and 'continuity' through the application of simple map-algebra techniques to time-slices (Lock and Daly 1999). A considerable body of work has been undertaken by Johnson in the context of the University of Sydney's TimeMap project. Here a sophisticated combination of commercial GIS and DBMS linked to bespoke software, has created a system for dynamically and interactively generating animation sequences from time-slice data. When coupled with the incorporation into the spatial database of explicitly temporal attributes such as chronological uncertainty (when a feature was located in a specific place) and diffuseness (how gradual change is from one condition to another), the result is a system that is taking the first steps towards the management and manipulation of spatio-temporal data (Johnson 1999).

As with the other technological developments discussed, interested readers are referred to Langran (1992) and Worboys (1995:302-313) for more detailed discussions of the practical mechanics of designing and implementing a TGIS.

12.7 TECHNOLOGICAL CONVERGENCE AND FIELD ARCHAEOLOGY

One of the most significant trends in technological development in the last 10 years or so has been *convergence*: the tendency for existing boundaries between systems to become blurred, and for new devices and categories of system to emerge. We explained in Chapter 1 that spatial technologies such as GIS are already a product of this kind of trend, being formed from a convergence of computer-aided mapping, database management systems and spatial analysis. However, it seems likely that the convergence of spatial technologies will continue beyond what we currently understand as GIS—in fact this is one of the principal reasons that we favour the term 'spatial technologies' over the more defined (and contested) 'Geographic Information Systems'.

Archaeology is now familiar with quite a range of spatial computer technologies, each with distinct applications to archaeological data processing and management. These include Database management systems, Raster-based systems *e.g.* image processing, aerial photography, geophysics, Vector-based systems (CAD, CAC, COGO), Survey systems (total stations *etc.*), 3D modelling and 'virtual reality' systems. We are also now becoming comfortable with the breaking down of these technological boundaries, such that software systems increasingly do more than one of these elements:

- dbms/vector/raster/spatial analysis (GIS) ;
- Survey/CAD with image processing ('geomatics');
- CAD/3D VR ('data visualisation').

Things begin to get really interesting when we consider the recent trends in technological convergence which include both hardware and software. Important in these are developments in:

- very mobile computers such as palmtops and even 'wearable' computers;
- GPS (see Chapter 3);
- mobile telecommunications;
- digital imaging.

If we mentally converge these with already-converging software technologies, then we might start to think in terms of *mobile, context-aware* spatial systems that have significant implications for the way that we do archaeology.

These systems may be more significant than simply smaller versions of existing software. With a mobile computer, context awareness (which is partly, but not entirely, GPS position fixing) and appropriate 'converged' GIS/DBMS/3D software we can imagine a whole variety of new fieldwork methods.

Surveys and site visits may involve the field archaeologist recording notes and observations complete with their spatial relevance, cross references or hyperlinks—essentially encoding spatial and topological components of textual information. Palmtop computers that implement this will permit the incorporation of digital photography or video footage in the same record. Fieldwalking may become significantly less mechanistic: with a pre-prepared survey plan and context-aware recording device there will be no need to set out, plan or compass. Instead an 'avatar' will appears on the screen, on top of a map or aerial photograph, and guide the walker to the appropriate observation point. Once there, the record could be made 'on the spot' with no need to remove items. Those that require identification could be removed, or merely photographed or video recorded for later identification. Excavation should also benefit, because archaeologists can work with the full knowledge of any geophysical or fieldwalking results, graphically presented as trenches are set out or excavated.

These developments are far from the fanciful imaginings that they might have been only two or three years ago. Practical applications of these kinds of synergies between spatial information technologies are already taking place in archaeology (Ryan *et al.* 1999, Craig 2000, Ziebart *et al.* in press). However, the real implications of these new ways of working may go far beyond the level of increased fieldwork efficiency and may actually impact on the structure of archaeological fieldwork itself.

To justify this assertion, we need only to consider how archaeological fieldwork has changed steadily over the last decades. Certainly the process has

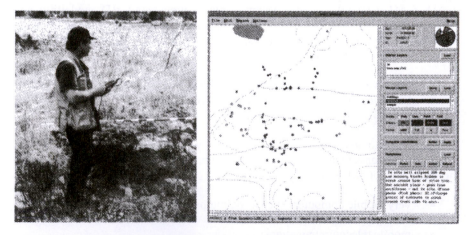

Figure 12.4 Field data acquisition with context-aware, mobile spatial systems. Illustrations are of the site of El Gandul, Andalucia and are taken from Ryan *et al.* (1999). Nick Ryan, Computing Laboratory, University of Kent, Canterbury. Reproduced with permission.

become more sophisticated and more complex, requiring increasing technical and methodological skill just to record information. This has led to the familiar, long post-excavation (or post fieldwork) period of analysis, and a very clear division between the (often mechanistic) fieldwork itself, and the highly interpretative post excavation work. In some instances, these have become fossilised into a distinction between archaeologists whose main aim is to recover and record material, and those whose job it is to make sense of them.

It is this distinction that these kinds of technologies may begin to dissolve. Tools themselves have no necessary effect on the empowerment or otherwise of fieldworkers, but these kinds of technologies are predicated on the provision of timely information that permits contingent fieldwork strategies—in other words an extreme shortening of the way in which interpretation feeds back to the decisions about where and what to observe, excavate or record. These kinds of ways of doing fieldwork are highly efficient. They will save money in many research and professional archaeological situations and so are very likely to be adopted. The really important thing is that they *depend on fieldworkers making sense of the information* and this is significant because it may begin to return the analytical responsibility, and the sense of engagement with the past, to fieldwork.

12.8 BUILDING A RESEARCH COMMUNITY

One characteristic of the growth of GIS has been the remarkable level of dynamism and proactive support offered by the GIS research community. Take for example the popular GIS packages ArcView. It currently offers the user a dedicated discussion list for airing and finding solutions to problems, along with a continually expanding library of useful scripts and extensions which serve to greatly enhance the functionality of the software (www.esri.com). Already archaeologists using GIS have an internet forum for the sharing of ideas (the gisarch mailing list) and have developed web-based guidance documents and suggestions for good practice (*e.g.* Gillings and Wise 1998, Bewley *et al.* 1999). These and others are available on the internet at http://www.ads.ahds.ac.uk/.

Possibly the greatest challenge for the future will be for the archaeological community to develop and enhance these mechanisms both through developing and sharing its own extensions, enhancements and solutions but, more importantly, by sharing knowledge and experience between all areas of the discipline—academic, fieldworker and cultural resource manager. If so, then those very categories may begin to dissolve themselves, and spatial technologies will enjoy a long and fruitful use-life within archaeology.

References

Abbott, E.A., 1932, *Flatland: a romance in many dimensions,* 4th edn., (London: Blackwell).

Adam, B., 1994, Perceptions of time. In *Companion encyclopedia of anthropology,* edited by Ingold, T., (London: Routledge), pp. 503-526.

Albrecht, J., 1996, *Universal GIS Operators - A task oriented systematization of data structure - independent GIS functionality leading towards a geographic modelling language,* Report, (Germany: ISPA Vechta).

Aldenderfer, M., 1996, Introduction. In *Anthropology, space and geographic information systems,* edited by Aldenderfer, M. and Maschner, H.D.G., Spatial information series (New York: Oxford University Press), pp. 3-18.

Aldenderfer, M. and Maschner, H.D.G. (eds.), 1996, *Anthropology, space and geographic information systems,* Spatial information series, (New York: Oxford University Press).

Allen, K.M.S., 1990, Modelling early historic trade in the eastern Great Lakes using geographic information. In Interpreting space: GIS and archaeology, edited by Allen, K.M.S., Green, S.W. and Zubrow, E.B.W., Applications of Geographic Information Systems (London: Taylor & Francis), pp. 319-329.

Allen, K.M.S., Green, S. and Zubrow, E.B.W. (eds.), 1990, Interpreting space: GIS and archaeology, Applications of Geographic Information Systems (London: Taylor & Francis).

Allinson, J., 1994, Proceedings of AGI conference, 1994, (Association for Geographic Information)

Amores, F., Garcia, L., Hurtado, V. and Rodriguez-Bobada, M.C., 1999, Geograhic information systems and archaeological resource management in Andalusia, Spain. In *New techniques for old times: CAA98 computer applications and quantitative methods in archaeology,* edited by Barceló, J.A., Briz, I. and Vila, A., (Oxford: Archaeopress), pp. 351-358.

Arnold III, J.B., 1979, Archaeological applications of computer drawn contour and three-dimensional perspective plots. In *Computer graphics in archaeology,* edited by Upham, S., Arizona University Anthropological Research Papers (Arizona: Arizona University), pp. 1-15.

Bailey, T.C., 1994, A review of statistical spatial analysis in geographic information systems. In *Spatial analysis and GIS,* edited by Fotheringham, S. and Rogerson, P., (London: Taylor & Francis), pp. 13-44.

Bailey, T.C. and Gatrell, A.C., 1995, *Interactive spatial data analysis,* (Harlow: Longman Scientific and Technical).

Barrett, J., 1994, *Fragments from antiquity,* (Oxford: Blackwells).

Basso, K.H., 1996, Wisdom sites in places: notes on a western Apache landscape. In *Senses of place,* edited by Feld, S. and Basso, K.H., (Santa Fe: School of American Research Advanced Seminar Series), pp. 53-90.

Baxter, M.J., 1994, *Exploratory multivariate analysis in archeology,* (Edinburgh: Edinburgh University Press).

Baxter, M.J., Beardah, C.C. and Wright, R.V.S., 1995, Some archaeological applications of kernel density estimates. *Journal of Archaeological Science,* **24**, pp. 347-354.

Beardah, C.C., 1999, Uses of multivariate kernel density estimates. In *Archaeology in the age of the internet: computer applications and quantitative methods in archaeology 1997*, edited by Dingwall, L., Exon, S., Gaffney, C.F., Laflin, S. and van Leusen, M., BAR International Series S750 (Oxford: Archaeopress), p. 107 [CD ROM].

Belcher, M., Harrison, A. and Stoddart, S., 1999, Analyzing Rome's hinterland. In *Geographic information systems and landscape archaeology*, edited by Gillings, M., Mattingly, D. and van Dalen, J., The archaeology of Mediterranean landscapes (Oxford: Oxbow Books), pp. 95-101.

Bell, T. and Lock, G.R., 2000, Topographic and cultural influences on walking the Ridgeway in later prehistoric times. In *Beyond the map: archaeology and spatial technologies*, edited by Lock, G.R., NATO Science Series A: Life Sciences 321, (Amsterdam: IOS Press), pp. 85-100.

Bell, T.L. and Church, R., 1985, Location-allocation modelling in archaeological settlement pattern research: some preliminary applications. *World Archaeology,* **16** (3), pp. 354-371.

Bettess, F., 1998, *Surveying for archaeologists,* 3rd edn., (Durham: University of Durham Department of Archaeology).

Bewley, R., Donaghue, D., Gaffney, V., van Leusen, M. and Wise, A., 1999, *Archiving aerial photography and remote sensing data,* (Oxford: Oxbow).

Bintliff, J.L., 1977, *Natural environment and human settlement in prehistoric Greece,* (Oxford: British Archaeological Reports).

Bintliff, J.L., 1996, The archaeological survey of the valley of the muses and its significance for Boeotian history. In *La Montagne des Muses*, edited by Hurst, A. and Schachter, A., (Geneva: Librairie Droz), pp. 193-219.

Bintliff, J.L. and Snodgrass, A.M., 1985, The Cambridge/Bradford Boeotia expedition: the first four years. *Journal of Field Archaeology,* **12**, pp.123-161.

Biswell, S., Cropper, L., Evans, J., Gaffney, C.F. and Leach, P., 1995, GIS and excavation: a cautionary tale from Shepton Mallet, Somerset, England. In *Archaeology and geographical information systems: a European perspective*, edited by Lock, G.R. and Stancic, Z., (London: Taylor & Francis), pp. 269-285.

Blasco, C., Baena, J. and Espiago, J., 1996, The role of GIS in the management of archaeological data: an example of application for the Spanish administration. In *Anthropology, space and geographic information systems*, edited by Aldenderfer, M.S. and Maschner, H.D.G., Spatial Information Series (New York: Oxford University Press), pp. 189-201.

Boaz, J.S. and Uleberg, E., 1993, Gardermoen project - use of a GIS system in antiquities registration and research. In *Computing the past: computer applications and quantitative methods in archaeology - CAA 92*, edited by Andresen, J., Madsen, T. and Scollar, I., (Aarhus: Aarhus University Press), pp. 177-182.

Bove, F.J., 1981, Trend surface analysis and the Lowland Classic Maya collapse. *American Antiquity,* **46** (1), pp. 93-112.

Bradley, R., 1991, Rock art and the perception of landscape. *Cambridge Archaeological Journal,* **1** (1), pp. 77-101.

Bradley, R., 1995, Making sense of prehistoric rock art. *British Archaeology*, (November 1995) p. 8.

Brandt, R., Groenewoudt, B.J. and Kvamme, K.L., 1992, An experiment in archaeological site location: modelling in the Netherlands using GIS techniques. *World Archaeology*, **24** (2), pp. 268-282.

Burrough, P.A., 1986, *Principles of GIS for land resources assessment*, (Oxford: Clarendon Press).

Burrough, P.A. and McDonnell, R.A., 1998, *Principles of geographic information systems*, (Oxford: Oxford University Press).

Butzer, K.W., 1982, *Archaeology as human ecology*, (Cambridge: Cambridge University Press).

Chadwick, A.J., 1978, A computer simulation of Mycenaean settlement. In *Simulation studies in archaeology*, edited by Hodder, I., (Cambridge: Cambridge University Press), pp. 47-57.

Chalmers, A., 1999, *What's this thing called science*, (Buckingham: Open University Press).

Chernoff, H., 1973, The use of faces to represent points in k-dimensional space graphically. *Journal of the American Statistical Association*, **68** (342), pp. 361-368.

Christaller, W., 1935, *Die zentralen orte in Suddeutcshland*, (Jena).

Christaller, W., 1966, *Central places in southern Germany; translated by Carlisle W. Baskin* (New Jersey: Prentice Hall).

Church, T., Brandon, R.J. and Burgett, G.R., 2000, GIS applications in archaeology: method in search of theory. In *Practical applications of GIS for archaeologists: a predictive modelling kit*, edited by Westcott, K.L. and Brandon, R.J., (London: Taylor & Francis), pp. 135-155.

Clark, G.A., Effland, R.W. and Johnstone, J.C., 1977, Proceedings of Computer Applications in Archaeology 1977, (Computer Centre, University of Birmingham).

Clarke, D.L., 1968, *Analytical archaeology*, (London: Methuen).

Clarke, D.L. (ed.) 1972, *Models in archaeology*, (London: Methuen).

Clarke, D.L., 1977a, *Spatial archaeology*, (London: Academic Press).

Clarke, D.L., 1977b, Spatial information in archaeology. In *Spatial archaeology*, edited by Clarke, D.L., (London: Academic Press), pp. 1-32.

Cleveland, W.S., 1993, *Visualising data*, (Murray Hill, N.J.: AT&T Bell Laboratories).

Cliff, A.D. and Ord, J.K., 1973, *Spatial autocorrelation*, (London: Pion).

Clubb, N.D. and Lang, N.A.R., 1996a, Learning from the acheivements of information sysems - the role of the post-implementation review in medium to large scale systems. In *Interfacing the past*, edited by Kamermans, H. and Fennema, K., Analecta Praehistorica Leidensia **28** (Leiden: Leiden University Press), pp. 73-80.

Clubb, N.D. and Lang, N.A.R., 1996b, A strategic appraisal of information systems for archaeology and architecture in England - past present and future. In *Interfacing the past*, edited by Kamermans, H. and Fennema, K., Analecta Praehistorica Leidensia **28** (Leiden: Leiden University Press), pp. 51-72.

COE, 1997, *Compendium of basic texts of the Council of Europe in the field of cultural heritage*, (Strasbourg: Council of Europe).

Cox, C., 1992, Satellite imagery, aerial photography and wetland archaeology: an interim report on an application of remote sensing to wetland archaeology: the pilot study in Cumbria, England. *World Archaeology,* **24** (2), pp. 249-267.

Craig, N., 2000, Real-time GIS construction and digital data recording of the Jiskairomuoko excavation, Peru. *SAA Bulletin,* **18** (1), (January 2000), pp. 24-28.

Crawford, O.G.S. and Keiller, A., 1928, *Wessex from the air,* (Oxford: Oxford University Press).

Cumming, P.A.B., 1997, *An assessment and enhancement of the sites and monuments record as a predictive tool for cultural resource management, development control and academic research,* PhD thesis, (Southampton: University of Southampton).

Cunliffe, B., 1971, Some aspects of Hillforts and their regional environments. In *The Iron Age and its Hillforts,* edited by Jesson, D.H.a.M., (Southampton: Southampton University Press), pp. 53-69.

Curry, M., 1998, *Digital places: living with geographic information technologies,* (London: Routledge).

Dalla Bona, L., 2000, Protecting cultural resources through forest management planning in Ontario using archaeological predictive modelling. In *Practical applications of GIS for archaeologists: a predictive modelling kit,* edited by Westcott, K.L. and Brandon, R.J., (London: Taylor & Francis), pp. 73-99.

Daly, P.T. and Lock, G.R., 1999, Timing is everything: commentary on managing temporal variables in geographic information ystems. In *New techniques for old times: CAA98 computer applications and quantitative methods in archaeology,* edited by Barceló, J.A., Briz, I. and Vila, A., BAR International Series S757 (Oxford: Archaeopress), pp. 287-293.

Date, C.J., 1990, *An introduction to database systems,* 5th edn., (New York: Addison Wesley).

DeMers, M.N., 1997, *Fundamentals of geographic information systems,* (New York: John Wiley & Sons).

Devereux, P., 1991, Three-dimensional aspects of apparent relationships between selected natural and artifcial features within the topography of the Avebury complex. *Antiquity,* **65,** pp. 894-898.

Douglas, D.H., 1994, Least-cost path in GIS using an accumulated cost surface and slopelines. *Cartographica,* **31** (Autumn 1994), pp.37-51.

Douglas, D.H., 1999 *Preface.* Web: http://aix1.uottawa.ca/~ddouglas/research/leastcos/cumcost4.htm, (Accessed 1999).

Duncan, R.B. and Beckman, K.A., 2000, The application of GIS predictive site location models within Pennsylvania and West Virginia. In *Practical applications of GIS for archaeologists: a predictive modelling kit,* edited by Westcott, K.L. and Brandon, R.J., (London: Taylor & Francis), pp. 33-58.

Eastman, J.R., 1993, *Idrisi Version 4.1 Update Manual,* (Worcester Massachusetts: Clark University).

Eastman, R.J., 1997, *Idrisi for Windows: user's guide version 2.0,* (Worcester Massachusetts: Clarke University).

Ebert, D., 1998, *Expanding the selection of tools for spatial analysis: geostatistics and the Als feildwalking data,* MSc Dissertation, (Southampton: University of Southampton).

Ebert, J.I., 2000, The state of the art in "inductive" predictive modelling: seven big mistakes (and lots of smaller ones). In *Practical applications of GIS for archaeologists: a predictive modelling kit*, edited by Westcott, K.L. and Brandon, R.J., (London: Taylor & Francis), pp. 129-134.

Ebert, J.I., Camilli, E.L. and Berman, M.J., 1996, GIS in the analysis of distributional archaeological data. In *New methods, old problems: geographic information systems in modern archaeological research*, edited by Maschner, H.D.G., (Carbondale, Illinois: Centre for Archaeological Investigations, Southern Illinois University at Carbondale), pp. 25-37.

Elmasri, R. and Navathe, S.B., 1989, *Fundamentals of database systems*, (New York: Addison Wesley).

Ericson, J. and Goldstein, R., 1980, Work space: a new approach to the analysis of energy expenditure within site catchments. In *Catchment Analysis: Essays on Prehistoric Resource Space*, edited by Findlow, F.J. and Ericson, J.E., Anthropology UCLA (Los Angeles: University of Califonia at Los Angeles), pp. 21-30.

ESRI, I., 1997 *Arc/Info Version 7.2.1 Online Help.*Web: (Accessed 2000).

Evans, C., 1985, Tradition and the cultural landscape: an archaeology of place. *Archaeological Review from Cambridge,* **4** (1), pp. 80-94.

Farley, J.A., Limp, W.F. and Lockhart, J., 1990, The archaeologists workbench: integrating GIS, remote sensing, EDA and database management. In *Interpreting space: GIS and archaeology*, edited by Allen, K.M.S., Green, S.W. and Zubrow, E.B.W., Applications of Geographic Information Systems (London: Taylor & Francis), pp. 141-164.

Feder, K.L., 1979, Geographic patterning of tool types as elicited by trend surface analysis. In *Computer graphics in archaeology*, edited by Upham, S., Arizona University Anthropological Research Papers (Arizona: Arizona University), pp. 95-102.

Findlow, F.J. and Ericson, J.E. (eds.), 1980, *Catchment analysis: essays on prehistoric resource space*, Anthropology UCLS, Volume 10, Numbers 1 & 2, (Los Angeles: University of California, Los Angeles).

Fisher, P., 1991, First experiments in viewshed uncertainty: the accuracy of the viewshed area. *Photogrammetric Engineering and Remote Sensing,* **57** (10), pp.1321-1327.

Fisher, P., 1992, First experiments in viewshed uncertainty: simulating fuzzy viewsheds. *Photogrammetric Engineering and Remote Sensing,* **58** (3), pp. 345-352.

Fisher, P., 1994, Probable and fuzzy models of the viewshed operation. In *Innovations in GIS*, edited by Worboys, M., (London: Taylor & Francis), pp. 161-175.

Fisher, P., 1995, An exploration of probable viewsheds in landscape planning. *Environment and Planning B: Planning and Design,* **22,** pp. 527-546.

Fisher, P., 1996a, Extending the applicability of viewsheds in landscape planning. *Photogrammetric Engineering and Remote Sensing,* **62** (11), pp. 527-546.

Fisher, P., 1996b, Reconsideration of the viewshed function in terrain modelling. *Geographical Systems,* **3,** pp. 33-58.

Fletcher, M. and Lock, G.R., 1980, Proceedings of *Computer applications in archaeology 1980*, (Computer Centre, University of Birmingham).

Fowler, M.J.F., 1995, High resolution Russian satellite imagery. *AARGnews: the newsletter of the Aerial Archaeology Research Group,* **11,** pp. 28-31.

Fowler, M.J.F., 1997, A cold war spy satellite image of Bury Hill, near Andover Hampshire. *HFCAS Newsletter,* **27,** pp. 5-7.

Fowler, M.J.F., 1999, A very high resolution satellite image of archaeological features in the vicinity of Mount Down, Hampshire. *HFCAS Newsletter,* **31,** pp. 4-5.

Fowler, M.J.F., 2000, In search of history. *Imaging Notes,* **15** (6), pp. 18-21.

Fowler, M.J.F., 2001 *Satellite remote sensing and archaeology homepage.* Web: http://ourworld.compuserve.com/homepages/mjff/homepage.htm, (Accessed 2001).

Fowler, M.J.F. and Curtis, H., 1995, Stonehenge from 230 kilometres. *AARGnews: the Newsletter of the Aerial Archaeology Research Group,* **11** (September 1995), pp. 8-15.

Fraser, D., 1983, *Land and society in neolithic orkney,* (Oxford: British Archaeological Reports).

Gaffney, V. and Stancic, Z., 1991, *GIS approaches to regional analysis: a case study of the island of Hvar,* (Ljubljana: Filozofska fakulteta).

Gaffney, V. and Stancic, Z., 1992, Didorus Siculus and the island of Hvar, Dalmatia: testing the text with GIS. In *Computer applications and quantitative methods in archaeology 1991,* edited by Lock, G.R. and Moffett, J., BAR International Series S577 (Oxford: Tempus Reparatum), pp. 113-125.

Gaffney, V. and van Leusen, M., 1995, Postscript - GIS, environmental determinism and archaeology: a parallel text. In *Archaeology and geographical information systems: a European perspective,* edited by Lock, G.R. and Stancic, Z., (London: Taylor & Francis), pp. 367-382.

Gaffney, V., Stancic, Z. and Watson, H., 1995, The impact of GIS on archaeology: a personal perspective. In *Archaeology and geographical information systems: a European perspective,* edited by Lock, G.R. and Stancic, Z., (London: Taylor & Francis), pp. 211-229.

Gaffney, V., Stancic, Z. and Watson, H., 1996, Moving from catchments to cognition: tentative steps towards a larger archaeological context for GIS. In *Anthropology, space and geographic information systems,* edited by Aldenderfer, M. and Maschner, H.D.G., Spatial information series (New York: Oxford University Press), pp. 132-154.

Gaffney, V., Ostir, K., Podobnikar, T. and Stancic, Z., 1996, Satellite imagery and GIS applications in Mediterranean landscapes. In *Interfacing the past,* edited by Kamermans, H. and Fennema, K., Analecta Praehistorica Leidensa 28, (Leiden: Leiden University Press), pp. 338-342.

Gamble, C., 2001, *Archaeology: the basics,* (London: Routledge).

Garcia-Sanjuan, L. and Wheatley, D., 1999, The state of the Arc: differential rates of adoption of GIS for European heritage management. *European Journal of Archaeology,* **2** (2), pp. 201-228.

Gell, A., 1992, *The anthropology of time: cultural constructions of temporal maps and images,* (Oxford: Berg).

Gillings, M., 1995, Flood dynamics and settlement in the Tisza valley of north-east Hungary: GIS and the Upper Tisza project. In *Archaeology and geographical information systems: a European perspective,* edited by Lock, G. and Stancic, Z., (London: Taylor & Francis), pp. 67-84.

Gillings, M., 1997, Not drowning, but waving? The Tisza flood-plain revisited. In *Archaeological applications of GIS: proceedings of Colloquium II, UISPP XIIIth Congress, Forli, Italy, September 1996*, edited by Johnson, I. and North, M., Archaeological Methods Series (Sydney: Sydney University), [CD ROM].

Gillings, M., 1998, Embracing uncertainty and challenging dualism in the GIS-based study of a palaeo-flood plain. *European Journal of Archaeology*, **1** (1), pp. 117-144.

Gillings, M., 2000, The utility of the GIS approach in the collection, management, storage and analysis of surface survey data. In *The future of surface artefact survey in Europe*, edited by Bintliff, J.L., Kuna, M. and Venclova, N., (Sheffield: Sheffield Academic Press), pp. 105-120.

Gillings, M. and Goodrick, G.T., 1996, Sensuous and reflexive GIS: exploring visualisation and VRML. *Internet Archaeology*, **1**.

Gillings, M. and Sbonias, K., 1999, Regional survey and GIS: the Boeotia project. In *Geographic information systems and landscape archaeology*, edited by Gillings, M., Mattingly, D. and van Dalen, J., The archaeology of Mediterranean landscapes (Oxford: Oxbow Books), pp. 35-54.

Gillings, M. and Wise, A.L., 1998, *GIS guide to good practice*, (Oxford: Oxbow).

Goodchild, M.F., 1996, Geographic information systems and spatial analysis in the social sciences. In *Anthropology, space and geographic information systems*, edited by Aldenderfer, M. and Maschner, H.D.G., Spatial Information Series (New York: Oxford University Press), pp. 241-250.

Gorenflo, L.J. and Gale, N., 1990, Mapping regional settlement in information space. *Journal of Anthropological Archaeology*, **9**, pp. 240-274.

Gosden, C., 1994, *Social being and time*, (Oxford: Blackwell).

Gosden, C., 1999, *Anthropology and archaeology: a changing relationship*, (London: Routledge).

Grant, E. (ed.) 1986a, *Central places, archaeology and history*, (Sheffield: University of Sheffield).

Grant, E., 1986b, Hill-forts, central places and territories. In *Central places, archaeology and history*, edited by Grant, E., (Sheffield: University of Sheffield), pp. 13-26.

Green, S., 1990, Approaching archaeological space: an introduction to the volume. In *Interpreting space: GIS and archaeology*, edited by Allen, K.M.S., Green, S.W. and Zubrow, E.B.W., Applications of Geographic Information Systems (London: Taylor & Francis), pp. 3-8.

Guillot, D. and Leroy, G., 1995, The use of GIS for archaeological resource management in France: the SCALA project, with a case study in Picardie. In *Archaeology and geographical information systems: a European perspective*, edited by Lock, G.R. and Stancic, Z., (London: Taylor & Francis), pp. 15-26.

Hageman, J.B. and Bennett, D.A., 2000, Construction of digital elevation models for archaeological applications. In *Practical applications of GIS for archaeologists: a predictive modelling kit*, edited by Westcott, K.L. and Brandon, R.J., (London: Taylor & Francis), pp. 113-127.

Haggett, P., 1965, *Locational analysis in geography* (London: Arnold).

Hagget, P. and Chorley, R., 1969, *Network analysis in geography*, (London: Edward Arnold).

Harris, T.M., 1985, GIS design for archaeological site information retrieval and predictive modelling. In *Professional archaeology in Sussex: the next five years*, (Proceedings of a one day conference by Association of Field Archaeology).

Harris, T.M., 1986, Proceedings of *Computer applications in archaeology 1986*, (Computer Centre, University of Birmingham).

Harris, T.M. and Lock, G.R., 1990, The diffusion of a new technology: a perspective on the adoption of geographic information systems within UK archaeology. In *Interpreting space: GIS and archaeology*, edited by Allen, K.M.S., Green, S.W. and Zubrow, E.B.W., Applications of Geographic Information Systems (London: Taylor & Francis),

Harris, T.M. and Lock, G.R., 1995, Toward an evaluation of GIS in European archaeology: the past, present and future of theory and applications. In *Archaeology and geographical information systems: a European perspective*, edited by Lock, G.R. and Stancic, Z., (London: Taylor & Francis), pp. 349-365.

Harris, T.M. and Lock, G.R., 1996, Multi-dimensional GIS exploratory approaches to spatial and temporal relationships within archaeological stratigraphy. In *Interfacing the past*, edited by Kamermans, H. and Fennema, K., Analecta Praehistorica Leidensia **28** (Leiden: Leiden University Press), pp. 307-316.

Hartwig, F. and Dearing, B.E., 1979, *Exploratory data analysis*, (Beverley-Hills: Sage).

Haselgrove, C., Millett, M. and Smith, I. (eds.), 1985, *Archaeology from the ploughsoil*, (Sheffield: Department of Archaeology and Prehistory, University of Sheffield).

Hillier, B. and Hanson, J., 1984, *The social logic of space*, (Cambridge: Cambridge University Press).

Hodder, I., 1972, Location models and the study of Romano-British Settlement. In *Models in archaeology*, edited by Clarke, D.L., (London: Methuen), pp. 887-909.

Hodder, I. (ed.) 1978, *Simulation studies in archaeology*, New directions in archaeology, (Cambridge: Cambridge University Press).

Hodder, I., 1992, *Theory and practice in archaeology*, (London: Routledge).

Hodder, I. and Orton, C., 1976, *Spatial analysis in archaeology*, (Cambridge: Cambridge University Press).

Hodgson, M.E., 1995, What cell size does the computed slope/aspect angle represent?. *Photogrammetric Engineering and Remote Sensing*, **61** (5), pp. 513-517.

Hunt, E.D., 1992, Upgrading site-catchment analyses with the use of GIS: investigating the settlement patterns of horticulturalists. *World Archaeology*, **24** (2), pp. 283-309.

Hutt, S., Jones, E.W. and McAllister, M.E., 1992, *Archaeological resource management*, (Washington D.C.: The Preservation Press).

Ingold, T., 1986, *The appropriation of nature: essays on human ecology and social relations*, (Manchester: Manchester University Press).

Isaaks, E.H. and Srivastava, R.M., 1989, *An introduction to applied geostatistics*, (Oxford: Oxford University Press).

Jain, R., 1980, Fuzzyism and real world problems. In *Fuzzy sets: theory and applications to policy analysis and information systems*, edited by Wang, P.P. and Chang, S.K., (New York: Plenum Press), pp. 129-132.

Johnson, I., 1999, Mapping the fourth dimension: the TimeMap project. In *Archaeology in the age of the internet: CAA97*, edited by Dingwall, L., Exon, S., Gaffney, V., Laflin, S. and van Leusen, M., BAR International Series S750 (Oxford: Archaeopress), p. 82 [CD-ROM].

Johnson, I. and North, M. (eds.), 1997, *Archaeological applications of GIS: proceedings of Colloquium II, UISPP XIIIth Congress, Forli, Italy, September 1996*, Archaeological Methods Series 5 [CD-ROM], (Sydney: Sydney University).

Johnson, S., 1989, *Hadrian's Wall,* (Oxford: English Heritage/Batsford).

Judge, W.J. and Sebastian, L. (eds.), 1988, *Quantifying the present and predicting the past: theory, method and application of archaeological Predictive modelling*, (Denver: US Department of the Interior, Bureau of Land Management).

Kamermans, H., 1999, Predictive modelling in Dutch archaeology, joining forces. In *New techniques for old times. CAA98 Computer applications and quantitative methods in archaeology*, edited by Barceló, J.A., Briz, I. and Vila, A., (Oxford: Archaeopress), pp. 225-229.

Kamon, E., 1970, Negative and positive work in climbing a laddermill. *Journal of Applied Physiology,* **29**, pp. 1-5.

Kohler, T.A., 1988, Predictive locational modelling: history and current practice. In *Quantifying the present, predicting the past*, edited by Judge, W.J. and Sebastian, L., (Denver, Colorado: US Department of the Interior Bureau of Land Management), pp. 19-59.

Kuna, M., 2000, Session 3 discussion: comments on archaeological prediction. In *Beyond the map: archaeology and spatial technologies* edited by Lock, G., NATO Science Series A: Life Sciences (Amsterdam: IOS Press), pp. 180-186.

Kvamme, K., 1983a, Computer processing techniques for regional modelling of archaeological site locations. *Advances in Computer Archaeology,* **1**, pp. 26-52.

Kvamme, K.L., 1983b, *A manual for predictive site location models: examples from the Grand Junction District, Colorado,* (Colorado: Bureau of Land Management, Grand Junction District).

Kvamme, K.L., 1988, Using existing archaeological survey data for model building. In *Quantifying the present, predicting the past*, edited by Judge, W.J. and Sebastian, L., (Denver, Colorado: US Department of the Interior Bureau of Land Management), pp. 301-428.

Kvamme, K.L., 1990a, The fundamental principles and practice of predictive archaeological modelling. In *Mathematics and information science in archaeology: a flexible framework*, edited by Voorips, A., Studies in Modern Archaeology (Bonn: Holos-Verlag), pp. 257-295.

Kvamme, K.L., 1990b, GIS algorithms and their effects on regional archaeological analyses. In *Interpreting space: GIS and archaeology* edited by Allen, K.M.S., Green, S.W. and Zubrow, E.B.W., Applications of Geographic Information Systems (London: Taylor & Francis) pp. 112-125.

Kvamme, K.L., 1990c, One-sample tests in regional archaeological analysis: new possibilities through computer technology. *American Antiquity,* **55** (2), pp.367-381.

Kvamme, K.L., 1990d, Spatial autocorrelation and the Classic Maya Collapse revisited: refined techniques and new conclusions. *Journal of Archaeological Science,* **17**, pp. 197-207.

Kvamme, K.L., 1993, Spatial analysis and GIS: an integrated approach. In *Computing the past: computer applications and quantitative methods in archaeology - CAA92*, edited by Andresen, J., Madsen, T. and Scollar, I., (Aarhus: Aarhus University Press), pp. 91-103.

Kvamme, K.L., 1997, Ranters corner: bringing the camps together: GIS and ED. *Archaeological Computing Newsletter,* **47** (Spring 1997), pp. 1-5.

Lake, M.W., 2000, Computer simulation of mesolithic foraging. In *Dynamics in human and primate societies: agent-based modelling of social and spatial processes*, edited by Gumerman, G. and Kohler, T.A., (Oxford: Oxford University Press),

Lake, M.W., In Press, MAGICAL computer simulation of mesolithic foraging on Islay. In *Hunter-gatherer landscape archaeology: the Southern Hebrides Mesolithic Project, 1988-1998*, edited by Mithen, S.J., (Cambridge: The MacDonald Institute for Archaeological Research).

Lake, M.W., Woodman, P.E. and Mithen, S.J., 1998, Tailoring GIS software for archaeological applications: an example concerning viewshed analysis. *Journal of Archaeological Science,* **25**, pp. 27-38.

Lang, N.A.R., 1993, From model to machine: procurement and implementation of Geographical Information Systems for county sites and monuments records. In *Computing the past: computer applications and quantitative methods in archaeology - CAA 92*, edited by Andresen, J., Madsen, T. and Scollar, I., (Aarhus: Aarhus University Press), pp. 167-176.

Langran, G., 1992, *Time in geographic information systems,* (New York: Taylor & Francis).

Leonard, R.D. and Jones, G.T. (eds), 1989, *Quantifying diversity in archaeology*, New Directions in Archaeology, (Cambridge: Cambridge University Press).

Lillesand, T.M. and Kiefer, R.W., 1994, *Remote sensing and image interpretation,* (New York: John Wiley & Sons).

Llobera, M., 1996, Exploring the topography of mind: GIS, social space and archaeology. *Antiquity,* **70**, pp. 612-622.

Llobera, M., 1999, Understanding movement: a pilot model towards the sociology of movement. In *Beyond the map: archaeology and spatial technologies*, edited by Lock, G.R., (Amsterdam: IOS Press), pp. 65-84.

Lock, G.R., Bell, T. and Lloyd, J., 1999, Towards a methodology for modelling surface survey data: the Sangro Valley project. In *Geographic information systems and landscape archaeology*, edited by Gillings, M., Mattingly, D. and van Dalen, J., The archaeology of Mediterranean landscapes NATO Science Series A: Life Sciences, (Oxford: Oxbow Books), pp. 55-63.

Lock, G.R. (ed.) 2000, *Beyond the map: archaeology and spatial technologies*, NATO Science Series A: Life Sciences, (Amsterdam: IOS Press).

Lock, G.R. and Daly, P.T., 1999, Looking at change, continuity and time in GIS: an example from the Sangro Valley, Italy. In *New techniques for old times: CAA98 computer applications and quantitative methods in archaeology*, edited by Barceló, J.A., Briz, I. and Vila, A., BAR International Series S757 (Oxford: Archaeopress), pp. 259-263.

Lock, G.R. and Harris, T.M., 1991, Integrating spatial information in computerised Sites and Monuments Records: meeting archaeological requirements in the 1990s. In *Computer applications and quatitative methods in archaeology 1990,* edited by Lockyear, K. and Rahtz, S.P.Q., BAR International Series S565 (Oxford: British Archaeological Reports), pp. 165-173.

Lock, G.R. and Harris, T.M., 1996, Danebury revisited: an English Iron Age hillfort in a digital landscape. In *Anthropology, space and geographic information systems* edited by Aldenderfer, M. and Maschner, H.D.G., Spatial Information Series (New York: Oxford University Press), pp. 214-240.

Lock, G.R. and Stancic, Z. (eds.), 1995, *Archaeology and geographical information systems: a European perspective,* (London: Taylor & Francis).

Loots, L., 1997, The use of projective and reflective viewsheds in the analysis of the Hellenistic defence system at Sagalassos. *Archaeological Computing Newsletter,* **49**, pp. 12-16.

Loots, L., Knackaerts, K. and Waelkens, M., 1999, Fuzzy viewshed analysis of the Hellenistic city defence system at Sagalassos, Turkey. In *Archaeology in the age of the internet: CAA97,* edited by Dingwall, L., Exon, S., Gaffney, V., Laflin, S. and van Leusen, M., BAR International Series S750 (Oxford: Archaeopress), pp. 82 [CD-ROM].

Mackie, Q., 1998, *The archaeology of Fjordland archipelagos: mobility networks, social practice and the built environment,* PhD thesis, (Southampton: University of Southampton).

Madrey, S.L.H. and Crumley, C.L., 1990, An application of remote sensing and GIS in a regional archaeological settlement pattern analysis: the Arroux River valley, Burgundy, France. In *Interpreting space: GIS and archaeology,* edited by Allen, K.M.S., Green, S.W. and Zubrow, E.B.W., Applications of Geographic Information Systems (London: Taylor & Francis), pp. 364-380.

Marble, D.F., 1990, The potential methodological impact of geographic information systems on the social sciences. In *Interpreting space: GIS and archaeology,* edited by Allen, K.M.S., Green, S.W. and Zubrow, E.B.W., Applications of Geographic Information Systems (London: Taylor & Francis),

Marble, D.F., 1996, *The human effort involved in movement over natural terrain: a working bibliography,* Report, (Ohio: Department of Geography, Ohio State University).

Maschner, H.D.G. (ed.) 1996a, *New methods, old problems: geographical information systems in modern archaeological research,* Occasional Paper No. 23, (Carbondale: Center for Archaeological Investigations, Southern Illinois University at Carbondale).

Maschner, H.D.G., 1996b, The politics of settlement choice on the Northwest coast cognition, GIS, and coastal landscapes. In *Anthropology, space and geographic information systems,* edited by Aldenderfer, M.S. and Maschner, H.D.G., Spatial Information Series (New York: Oxford University Press), pp. 175-189.

Matheron, G., 1971, *The theory of regionalised variables and its applications,* (Paris: Ecole Nationale Superieure des Mines de Paris).

McGrew, J.C. and Monroe, C.B., 1993, *An introduction to statistical problem solving in geography,* (Oxford: Wm. C. Brown Publishers).

Miller, A.P., 1996, Digging deep: GIS in the city. In *Interfacing the past,* edited by Kamermans, H. and Fennema, K., Analecta Praehistorica Leidensa **28** (Leiden: Leiden University Press), pp. 369-376.

Minetti, A.E., 1995, Optimum gradient of mountain paths. *Journal of Applied Physiology,* **79** (5), pp. 1698-1703.

Murray, D.M., 1995, The management of archaeological information: a strategy. In *Computer applications and quantitative methods in archaeology 1993*, edited by Wilcock, J. and Lockyear, K., BAR International Series S598 (Oxford: Tempus Reparatum), pp. 83-87.

Mytum, H., 1996, Intrasite patterning and the temporal dimension using GIS: the example of Kellington churchyard. In *Interfacing the past*, edited by Kamermans, H. and Fennema, K., Analecta Praehistorica Leidensa **28** (Leiden: Leiden University Press), pp. 275-292.

Neiman, F.D., 1997, Conspicuous consumption as wasteful advertising: a Darwinian perspective on spatial patterns in Classic Maya terminal monuments dates. In *Rediscovering Darwin: Evolutionary theory and archaeological explanation*, edited by Barton, M.C. and Clark, G.A., (Archaeological Papers of the American Anthropological Association), pp. 267-290.

Openshaw, S. and Openshaw, C., 1997, *Artificial intelligence in geography*, (New York: John Wiley & Sons).

Palumbo, G., 1993, JADIS (Jordan Antiquities Database and Information System): an example of national archaeological inventory and GIS applications. In *Computing the past: computer applications and quantitative methods in archaeology - CAA 92*, edited by Andresen, J., Madsen, T. and Scollar, I., (Aarhus: Aarhus University Press), pp. 183-194.

Peucker, T.K., Fowler, R.J., Little, J. J. and Mark, D.M., 1978, *Proceedings of the ASP Digital Terrain Models (DTM) Symposium*, 1978, (Bethesda, Maryland: American Society of Photogrammetry).

Peuquet, D.J., 1977, *Raster data handling in geographic information systems*, (Buffalo: Geographic Information Systems Laboratory, New York State University).

Pickles, J., 1995, *Ground truth: the social implications of geographic information systems*, (New York: Guilford Press).

Pitt-Rivers, A., 1892, *Excavations at Bokerly and Wansdyke*.

Rahtzel, F., 1882, *Anthropogeographie, oder Grundzuge der Anwendung der Erdlkunde auf die Geschichte*, (Stuttgart: Engelhorn).

Raper, J. (ed.) 1989, *Three dimensional applications in geographical information systems*, (London: Taylor & Francis).

Renfrew, C., 1973, Monuments, mobilization and social organization in Neolithic Wessex. In *The explanation of culture change: models in prehistory*, edited by Renfrew, C., (London: Duckworth), pp. 539-558.

Renfrew, C., 1979, *Investigations in Orkney*, (London: Society of Antiquaries).

Robinson, J.M. and Zubrow, E., 1999, Between spaces: interpolation in archaeology. In *Geographic information systems and landscape archaeology*, edited by Gillings, M., Mattingly, D. and van Dalen, J., The archaeology of Mediterranean landscapes (Oxford: Oxbow Books), pp. 65-83.

Roorder, I.M. and Wiemer, R., 1992, Towards a new archaeological information system in the Netherlands. In *Computer applications and quantitative methods in archaeology 1991*, edited by Lock, G. and Moffett, J., (Oxford: British Archaeological Reports), pp. 85-88.

Rosenfield, A., 1980, Tree structures for region representation. In *Map data processing*, edited by Freeman, H. and Pieroni, G.G., (London: Academic Press), pp. 137-150.

Ruggles, C.L.N. and Church, R.L., 1996, Spatial allocation in archaeology: an opportunity for reevaluation. In *New methods, old problems: Geographic Information Systems in modern archaeological research*, edited by Maschner, H.D.G., (Carbondale, Illinois: Center for Archaeological Investigations, Southern Illinois University at Carbondale), pp. 147-173.

Ruggles, C.L.N. and Medyckyj-Scott, D.J., 1996, Site location, landscape visibility and symbolic astronomy: a Scottish case study. In *New methods, old problems: Geographic Information Systems in modern archaeological research*, edited by Maschner, H.D.G., (Carbondale, Illinois: Center for Archaeological Investigations, Southern Illinois University at Carbondale), pp. 127-146.

Ruggles, C.L.N., Martlew, R. and Hinge, P., 1991, The North Mull project (2): the wider astronomical significance of the sites, *Archaeoastronomy*, **16**, pp. 51-75.

Ruggles, C.L.N., Medyckyj-Scott, D.J. and Gruffydd, A., 1993, Multiple viewshed analysis using GIS and its archaeological application: a case study in northern Mull. In *Computing the past: computer applications and quantitative methods in archaeology - CAA92*, edited by Andresen, J., Madsen, T. and Scollar, I., (Aarhus: Aarhus University Press), pp. 125-132.

Ryan, N.S. and Smith, D.J., 1995, *Database systems engineering*, (London: International Thompson Computer Press).

Ryan, N.S., Pascoe, J. and Morse, D., 1999, FieldNote: extending a GIS into the field. In *New techniques for old times: CAA98 computer applications and quantitative methods in archaeology*, edited by Barceló, J.A., Briz, I. and Vila, A., BAR International Series S757 (Oxford: Archaeopress), pp. 127-132.

Savage, S.H., 1990, Modelling the Late Archaic social landscape. In *Interpreting space: GIS and Archaeology*, edited by Allen, K.M.S., Green, S. and Zubrow, E.B.W., Applications of Geographic Information Systems, (London: Taylor and Francis), pp. 330-335.

Schofield, W., 1993, *Engineering surveying*, 4th edn., (London: Butterworth-Heinemann).

Schollar, I., Tabbagh, A., Hesse, A. and Herzog, I., 1990, *Archaeological prospecting and remote sensing*, 1st edn., (Cambridge: Cambridge University Press).

Shanks, M. and Tilley, C., 1987, *Social theory and archaeology*, (Cambridge: Polity Press).

Shapiro, M., 1993, *Grass 4.1 man page for s.surf.idw*.(US Corps of Engineers).

Shennan, S.J., 1997, *Quantifying archaeology*, 2nd edn., (Edinburgh: Edinburgh University Press).

Smith, N., 1995, Towards a study of ancient Greek landscapes: the Perseus GIS. In *Archaeology and geographical information systems: a European perspective*, edited by Lock, G. and Stancic, Z., (London: Taylor & Francis), pp. 239-248.

Sonka, M., Hlavac, V. and Boyle, R., 1993, *Image processing, analysis and machine vision*, (London: Chapman and Hall).

Spikins, P., 1995, Virtual landscapes - GIS and lithic scatters. In *Lithics in Context*, edited by Schofield, A.J., Lithic Studies Society Occasional Paper 5 (London: Lithic Studies Society).

Star, J. and Estes, J., 1990, *Geographic information systems*, (New Jersey: Prentice-Hall).

StatSoft Inc., 2001 *Electronic statistics textbook*. Web: http://www.statsoft.com/textbook/stathome.html, (Accessed 2001).

Taylor, P.J. and Johnston, R.J., 1995, Geographic information systems and geography. In *Ground truth: the social implications of Geographic Information Systems*, edited by Pickles, J., (New York: The Guildford Press), pp. 51-67.

Thomas, J., 1993, The politics of vision and the archaeologies of landscape. In *Landscape, politics and perspectives*, edited by Bender, B., (Oxford: Berg), pp. 19-48.

Thomas, J., 1996, *Time, culture and identity*, (London: Routledge).

Thrower, J.W., 1972, *Maps and man: an examination of cartography in relation to culture and civilization*, (New Jersey: Prentice-Hall).

Tilley, C., 1993, Art, architecture, landscape [Neolithic Sweden]. In *Landscape, politics and perspectives* edited by Bender, B., (Oxford: Berg), pp. 49-84.

Tilley, C., 1994, *A phenomenology of landscape: places, paths and monuments*, 1st edn., (Oxford/Providence USA: Berg).

Tobler, W., 1970, A computer movie simulating urban growth in the Detroit region. *Economic Geography*, **46** (2), pp. 234-240.

Tobler, W. and Chen, Z., 1993, A quadtree for global information storage *Geographical Analysis*, **14** (4), pp. 360-371.

Tomlin, C.D. (ed.) 1983, *A map algebra*, Proceedings of the Harvard Computer Conference, 1983, 31st July - 4th August (Cambridge, Massachusetts: Harvard).

Tomlin, C.D., 1990, *Geographic information systems and cartographic modelling*, (Englewood Cliffs, NJ: Prentice-Hall).

Tomlinson, R.F., 1982, Panel discussion: technology alternatives and technology transfer. In *Computer assisted cartography and information processing: hope and realism*, edited by Douglas, D.H. and Boyle, A.R., (Ottawa: Canadian Cartographic Association), pp. 65-71.

Trigger, B., 1989, *A history of archaeological thought*, (Cambridge: Cambridge University Press).

Tschan, A.P., 1999, An introduction to object-oriented GIS in archaeology. In *New techniques for old times: CAA98 computer applications and quantitative methods in archaeology*, edited by Barceló, J.A., Briz, I. and Vila, A., BAR International Series S757 (Oxford: Archaeopress), pp. 303-316.

Tukey, J.W., 1977, *Exploratory data analysis*, (Philippines: Addison-Wesley Publishing Company).

Upham, S. (ed.) 1979, *Computer graphics in archaeology*, Arizona University Anthropological Research Papers, (Arizona: Arizona University).

van Leusen, M., 1993, Cartographic modelling in a cell-based GIS. In *Computing the past: computer applications and quantitative methods in archaeology – CAA 92*, edited by Andresen, J., Madsen, T. and Scollar, I., (Aarhus: Aarhus University Press), pp. 105-123.

van Leusen, M., 1999, Viewshed and cost surface analysis using GIS (Cartographic modelling in a cell-based GIS II). In *New techniques for old times: CAA98*, edited by Barceló, J.A., Briz, I. and Vila, A., British Archaeological Reports International Series 757 (Oxford: Archaeopress), pp. 215-223.

Vita-Finzi, C. and Higgs, E., 1970, Prehistoric economy in the Mount Carmel area of Palestine: site catchemnt analysis. *Proceedings of the Prehistoric Society*, **36**, pp. 1-37.

von Thunen, J.H., 1966, *Der Isolierte Staat*, (Oxford: Pergamon).

Vullo, N., Fontana, F. and Guerreschi, A., 1999, The application of GIS to intra-site spatial analysis: preliminary results from Alpe Veglia (VB) and Mondeval de Sora (BL), two Mesolithic sites in the Italian Alps. In *New techniques for old times: CAA98*, edited by Barceló, J.A., Briz, I. and Vila, A., British Archaeological Reports International Series 757 (Oxford: Archaeopress), pp. 111-116.

Wadsworth, R. and Treweek, J., 1999, *Geographical information systems for ecology: an introduction*, (Harlow: Addison Wesley Longman Limited).

Walker, R. (ed.) 1993, *AGI Standards Committee GIS Dictionary*, (Edinburgh: Association for Geographic Information).

Wansleeben, M., 1988, Applications of geographical information systems in archaeological research. In *Computer and quantitative methods in archaeology 1988*, edited by Rahtz, S.P.Q., BAR International Series S446 (ii) (Oxford: British Archaeological Reports), pp. 435-451.

Wansleeben, M. and Verhart, L.B.M., 1995, GIS on different spatial levels and the neolithization process in the south-eastern Netherlands. In *Archaeology and geographical information systems: a European perspective*, edited by Lock, G. and Stancic, Z., (London: Taylor & Francis), pp. 153-169.

Warren, R.E. and Asch, D.L., 2000, A predictive model of archaeological site location in the eastern Prairie Peninsula. In *Practical applications of GIS for archaeologists: a predictive modelling kit*, edited by Westcott, K.L. and Brandon, R.J., (London: Taylor & Francis), pp. 5-32.

Wasserman, S. and Faust, K., 1994, *Social network analysis: methods and applications*, (Cambridge: Cambridge University Press).

Westcott, K.L. and Brandon, R.J. (eds), 2000, *Practical applications of GIS for archaeologists: a predictive modelling kit*, (London: Taylor & Francis).

Westcott, K.L. and Kuiper, J.A., 2000, Using a GIS to model prehistoric site distributions in the Upper Chesapeake Bay. In *Practical applications of GIS for archaeologists: a predictive modelling kit*, edited by Westcott, K.L. and Brandon, R.J., (London: Taylor & Francis), pp. 59-72.

Wheatley, D., 1993, Going over old ground: GIS, archaeological theory and the act of perception. In *Computing the past: computer applications and quantitative methods in archaeology - CAA 92*, edited by Andresen, J., Madsen, T. and Scollar, I., (Aarhus: Aarhus University Press), pp. 133-138.

Wheatley, D., 1995a, Cumulative viewshed analysis: a GIS-based method for investigating intervisibility, and its archaeological application. In *Archaeology and geographic information systems: a European perspective*, edited by Lock, G. and Stancic, Z., (London: Taylor & Francis), pp. 171-185.

Wheatley, D., 1995b, The impact of geographic information systems on the practice of archaeological management. In *Managing archaeology* edited by Cooper, M., Firth, A.J., Carmen, J. and Wheatley, D., Theoretical Archaeology Group Series (London: Routledge), pp. 163-174.

Wheatley, D., 1996a, Between the lines: the role of GIS-based predictive modelling in the interpretation of extensive field survey. In *Interfacing the past*, edited by Kamermans, H. and Fennema, K., Analecta Praehistorica Leidensa 28, (Leiden: Leiden University Press), pp. 275-292.

Wheatley, D., 1996b, The use of GIS to understand regional variation in Neolithic Wessex. In *New methods, old problems: Geographic Information Systems in modern archaeological research*, edited by Maschner, H.D.G., Southern Illinois University at Carbondale Occasional Paper No. 23, (Carbondale, Illinois:

Center for Archaeological Investigations, Southern Illinois University at Carbondale), pp. 75-103.

Wheatley, D., 1998, Ranters corner: keeping the camp fires burning: the case for pluralism, *Archaeological Computing Newsletter,* **50** (Spring 1998), 2-7.

Wheatley, D., 2000, Spatial technology and archaeological theory revisited. In *CAA96 computer applications and quantitative methods in archaeology,* edited by Lockyear, K., Sly, T.J.T. and Mihailescu-Birliba, V., BAR International Series S845 (Oxford: Archaeopress), pp. 123-131.

Wheatley, D. and Gillings, M., 2000, Vision, perception and GIS: developing enriched approaches to the study of archaeological visibility. In *Beyond the map: archaeology and spatial technologies,* edited by Lock, G.R., NATO Science Series A: Life Sciences (Amsterdam: IOS Press), pp. 1-27.

Whiteley, D.S. and Clark, W.A., 1985, Spatial autocorrelation tests and the Classic Maya collapse: methods and inferences. *Journal of Archaeological Science,* **12** pp. 377-395.

Whittle, A., Pollard, J. and Grigson, C., 1999, *The harmony of symbols: the Windmill Hill causewayed enclosure,* (Oxford: Oxbow books).

Williams, I., Limp, W.F. and Briuer, F.L., 1990, Using geographic information systems and exploratory data analysis for archaeological site classification and analysis. In *Interpreting space: GIS and archaeology,* edited by Allen, K.M.S., Green, S.W. and Zubrow, E.B.W., Applications of Geographic Information Systems, (London: Taylor & Francis), pp. 239-273.

Wilson, R.J., 1985, *Introduction to graph theory,* 3rd edn., (Harlow: Longman Scientific and Technical).

Wise, A.L., 2000, Building theory into GIS based landscape analysis. In *CAA96 computer applications and quantitative methods in archaeology,* edited by Lockyear, K., Sly, T.J.T. and Mihailescu-Birliba, V., BAR International Series S845 (Oxford: Archaeopress), pp. 141-147.

Woodman, P.E. and Woodward, M., in press, The use and abuse of statistical methods in archaeological site location modelling. In *Contemporary themes in archaeological computing,* edited by Wheatley, D., Earl, G. and Poppy, S., University of Southampton Department of Archaeology Monograph Series No. 3 (Oxford: Oxbow Books).

Worboys, M.F., 1995, *GIS: a computing perspective,* (London: Taylor & Francis).

Wynn, J.C., 1998 *A review of geophysical methods used in archaeology.*Web: http://www.terraplus.com/papers/wynn.htm, (Accessed 2001).

Zadeh, L.A., 1980, Foreword. In *Fuzzy sets: theory and applications to policy analysis and information systems,* edited by Wang, P.P. and Chang, S.K., (New York: Plenum Press).

Ziebart, M., Holder, N. and Dare, P., in press, Field digital data acquisition (FDA) using total station and pencomputer: a working methodology. In *Contemporary themes in archaeological computing,* edited by Wheatley, D., Earl, G. and Poppy, S., University of Southampton Department of Archaeology Monograph Series No. 3 (Oxford: Oxbow Books).

Zubrow, E.B.W., 1990a, Modelling and prediction with geographic information systems: a demographic example from prehistoric and historic New York. In *Interpreting space: GIS and archaeology,* edited by Allen, K.M.S., Green, S.W. and Zubro, E.B.W., Applications of Geographic Information Systems, (London: Taylor & Francis), pp. 307-318.

Zubrow, E.B.W., 1990b, Contemplating space: a commentary of theory. In *Interpreting space: GIS and archaeology*, edited by Allen, K.M.S., Green, S.W. and Zubrow, E.B.W., Applications of Geographic Information Systems, (London: Taylor & Francis), pp. 67-72.

Index

#0288 - 221116 - C0 - 229/152/16 - PB - 9780415246408